T0413942

Nanomaterials for Thermoelectric Devices

Nanomaterials for Thermoelectric Devices

edited by

Yong X. Gan

Published by

Pan Stanford Publishing Pte. Ltd.
Penthouse Level, Suntec Tower 3
8 Temasek Boulevard
Singapore 038988

Email: editorial@panstanford.com
Web: www.panstanford.com

British Library Cataloguing-in-Publication Data
A catalogue record for this book is available from the British Library.

Nanomaterials for Thermoelectric Devices

ISBN 978-981-4774-98-7 (Hardcover)
ISBN 978-0-429-48872-6 (eBook)

To

Feng, Bo, Ryan, Jeremy, and Kevin.

Contents

Preface

The objective of this book is to introduce thermoelectric energy conversion nanomaterials and the related manufacturing processes. With the increased demand for energy, we are facing the grand challenge of energy sustainability. Renewable energy catches great interest in solving the problem. Among the various renewable energy sources, solar heat and waste heat energy harvesting has significant advantages over others due to the availability. A high energy conversion efficiency is critical for practical application of thermoelectric energy conversion systems. To understand the fundamentals of energy conversion mechanisms is essential. Thermoelectric nanomaterials have an indispensable role in the conversion of heat to electricity energy. Therefore, it is necessary to introduce the latest research progress on thermoelectric energy conversion nanomaterials and manufacturing technology to readers at various levels.

The contents of the book deal with various nanomaterials and techniques for manufacturing thermoelectric energy conversion materials. A comprehensive state-of-the-art review on nanomaterials, related processing technologies, and applications is provided. Emphasis is put on the technological aspects of thermoelectric energy conversion materials and processing and manufacturing. Four aspects are discussed in the book: (i) the scope, goals, and fundamentals of energy harvesting and conversions, (ii) energy conversion material synthesis, processing, and structure characterization methods with emphasis on thermoelectric energy conversion nanomaterials, (iii) thermoelectric energy conversion device concepts, and (iv) nanomaterials processing and manufacturing techniques for thermoelectric energy conversions, including chemical, electrochemical, and external field–assisted manufacturing methods. The book also briefly presents thermoelectric material applications. The research results of various thermoelectric energy conversion materials processing and manufacturing are presented as well.

Chapter 1 is on the concept and fundamentals of thermoelectricity. Chapter 2 deals with micro- and nanoscale device concepts. Various micro- and nanofabrication and manufacturing techniques are reviewed. Chapter 3 focuses on thermoelectric silicon single-crystal nanowires made by self-catalyzed chemical etching. Electrochemical deposition of Bi-Te nanoparticles onto the Si nanowires is also introduced. The thermoelectric properties of silicon nanowires and Bi-Te-capped Si nanowires are compared.

In Chapter 4, the electrodeposited Te-Bi-Pb thermoelectric film is presented. A porous substrate made from electrochemical de-alloying zinc from brass was used to electroplate a thermoelectric Te-Bi-Pb alloy film. Results of the Seebeck effect of the film are given.

Chapter 5 is an overview of conducting polymer materials and their composites for thermoelectric energy conversion. Typical conducting polymers and the related processing methodologies are briefly reviewed. A case study of the thermoelectric property of a polyaniline (PANI)-coated TiO_2 nanotube nanocomposite is introduced. The conducting polymer, PANI, was prepared by electrochemical oxidation of the nanotubes. The morphology of the nanocomposites was observed. The highest absolute value of the Seebeck coefficient for the TiO_2 nanotube/polyaniline nanocomposite materials was revealed.

Chapter 6 introduces how to manufacture flexible thermoelectric materials. A bismuth telluride microscale particle–filled silicone rubber composite was prepared into millimeter-sized wires using the electrospinning approach. The composite wires were tested in view of the electrical resistance and Seebeck coefficient. Results are presented in this chapter.

Chapter 7 is on complex thermoelectric materials made by catalyst-assisted metal organic chemical vapor deposition. How to deposit Bi-Te-Ni-Cu-Au complex materials on an anodic aluminium oxide nanoporous substrate is shown.

Chapter 8 deals with the thermoelectric effect of porous materials. Porous materials are supposed to have the property of slowing down the heat flow. When a temperature gradient exists across a porous thermoelectric material, an electrical voltage is induced due to the difference between the energy levels of the electrons on the hot side and the electrons on the cold side. The infiltration casting method was used for the formation of a bismuth-based alloy. The Seebeck coefficient was measured, and the thermoelectric figure of merit was estimated.

In Chapter 9, injection casting is introduced as a relatively new method for thermoelectric alloy processing. Advances in manufacturing semiamorphous composite materials containing thermoelectric nanoscale grains by the controlled injection casting method are presented. Recent research results of the thermoelectric behavior of the nanocomposite materials are also provided.

Chapter 10 is on nanocasting thermoelectric composite materials. Manufacturing multicomponent organic/inorganic composite materials containing nanotubes, nanofibers, and nanoparticles for energy conversions by nanocasting is introduced. The effect of combined electric, magnetic, and/or mechanical forces on the distribution of nanoscale particles, tubes, or fibers into matrix materials is discussed. A new nanocasting manufacturing process during which electromagnetic and mechanical actions coexist is also introduced. Characterization of the properties of the thermoelectric materials with multiple components and complex structures is performed.

The last chapter, Chapter 11, deals with additive manufacturing of thermoelectric energy conversion functional composite materials by integrating electrohydrodynamic manufacturing into 3D printing. The author is trying to answer several questions in the chapter. These questions are: (i) how to design and make a new manufacturing machine consisting of an electrohydrodynamic manufacturing system and a 3D printer, (ii) how to use the machine to manufacture composite materials with functional thermoelectric functional nanoparticles uniformly distributed in polymer fibers, and (iii) how to heat-treat the composites to convert the polymer fibers into partially carbonized fibers. In the rest of the chapter, the composite materials are shown with a significant thermoelectric response for thermal sensing and energy conversion. Scalable manufacturing of composite material mats by this new technology is briefly touched upon at the end of the chapter.

Yong X. Gan

2018

Acknowledgments

This work was supported by the National Science Foundation (NSF) under Grant Number CMMI-1333044. The scanning electron microscopic (SEM) images were made possible through the NSF MRI grant DMR-1429674. Mr. Anan Hamdan is appreciated for his assistance in SEM experiments. The transmission electron microscopic (TEM) research carried out in the Center for Functional Nanomaterials, Brookhaven National Laboratory, was supported by the US Department of Energy, Office of Basic Energy Sciences, under Contract No. DE-AC02-98CH10886. Dr. Lihua Zhang is appreciated for her assistance in TEM experiments. The author acknowledges the support by the California State Polytechnic University Pomona 2016–2017 and 2017–2018 Provost's Teacher-Scholar program, the 2016–2017 SPICE grant program, and the 2016–2017 RSCA grant program. Special thanks are also due to the author's former graduate research assistants, including but not limited to Dr. Lusheng Su at the University of Toledo, Mr. Bruce Y. Decker, Mr. Manuel Eshaghof, Ms. Ann Doris Chen, and Mr. David Rodriguez at the California State Polytechnic University Pomona, for their contributions to the research, as described in the book.

Chapter 1

Thermoelectric Energy Conversion Fundamentals

Energy conversion materials and systems are designed to extract various types of energy from the environment to provide sustainable power for sensors, actuators, electronics, heaters, etc., thus lowering the need for batteries or power generating devices. Examples of ambient energy sources are temperature gradients and fluctuations and mechanical motions, such as impacts and vibrations, which are otherwise wasted or harmful. Thermoelectric energy conversion mechanisms are based on the Seebeck effect, that is as electrons move from the high-temperature end to the low-temperature end of a conductor, a current is formed and thus electric power can be generated. Recently, studies have been carried out on nanostructured energy conversion materials. This chapter deals with the mechanism of thermoelectric energy conversion. The physical principles behind thermoelectricity will be discussed.

1.1 Thermoelectricity

1.1.1 Seebeck Effect and Thermoelectric Power

Thermoelectric (TE) energy conversion is based on the Seebeck effect, which was discovered in 1821 by the Estonian German

Nanomaterials for Thermoelectric Devices
Yong X. Gan

ISBN 978-981-4774-98-7 (Hardcover), 978-0-429-48872-6 (eBook)
www.panstanford.com

physicist Thomas Johann Seebeck. In this effect, a temperature difference across the TE materials, preferably a p-type and an n-type semiconductor, as shown in Fig. 1.1, generates electricity, that is $\Delta T \Rightarrow I$, where T stands for temperature and I represents electric current. The reverse effect is that electricity causes active cooling. Thermoelectricity may be simply demonstrated by using the installation shown in Fig. 1.2. In this demonstration, bismuth telluride alloy (Bi-Te) TE modules are installed between two aluminum legs. One leg is immersed in hot water, while the other is inserted into cold water. Due to the temperature difference, the Bi-Te unit generates electricity and drives the small electric fan to rotate.

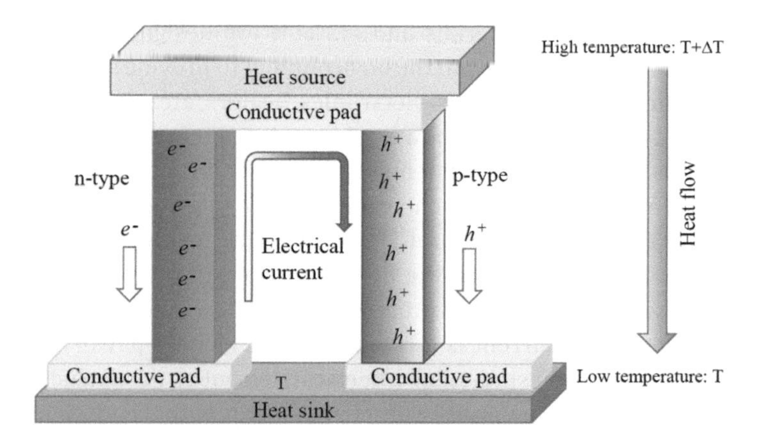

Figure 1.1 Illustrations of thermoelectric energy conversion.

TE phenomena can be described by the second law of thermodynamics. The second law can be stated in several ways. Lord Kelvin's statement of the second law is as follows: "A transformation whose only final result is to transform into work heat extracted from a source which is at the same temperature throughout is impossible." Since there is a temperature difference, some of the thermal energy from the heat source as shown in Fig. 1.1 is converted into work by the TE semiconducting materials, which is not in contradiction to the second law as stated by Kelvin.

The Seebeck effect can also be explained in terms of the entropy change. As known, the change in entropy is

$$\Delta S = \frac{Q}{T}, \tag{1.1}$$

where ΔS is the change in entropy, Q is the heat transferred, and T is the temperature at which heat is transferred.

If the heat transfer occurs only between the heat source and the heat sink, the change in the entropy of the high-temperature end (the heat source), ΔS_H, should be

$$\Delta S_H = \frac{Q_H}{T + \Delta T} < 0, \tag{1.2}$$

where Q_H is the transferred heat at the high-temperature end.

ΔS_H is negative because heat flows from the high-temperature end, the heat source, into the TE materials.

Figure 1.2 Installation showing thermoelectric energy conversion.

The change in the entropy of the low-temperature end (the heat sink), ΔS_L, should be

$$\Delta S_L = \frac{Q_L}{T} > 0, \tag{1.3}$$

where Q_L is the transferred heat at the low-temperature end.

ΔS_L is positive because heat transfers from the TE materials into the low-temperature end (heat sink).

According to the second law of thermodynamics, the total change in entropy, $\Delta S = \Delta S_H + \Delta S_L$, must be positive. In other words, the process takes place spontaneously when $\Delta S_H < \Delta S_L$, that is,

$$\frac{Q_H}{T + \Delta T} < \frac{Q_L}{T}. \tag{1.4}$$

During the TE energy conversion process, a portion of the heat from the heat source is converted into electrical energy and will not be rejected into the heat sink, which means that $Q_H > Q_L$. This comes from the requirement of conservation of energy, or the first law of thermodynamics. For Eq. 1.4 to be true, ΔT must be nonzero. Therefore, a temperature difference is the driving force for TE energy conversion, as shown by the illustration of Fig. 1.1.

The Seebeck effect can be explained by solid-state physics. If a temperature difference exists in the TE materials (semiconductors), as shown in Fig. 1.1, the heat entering the module raises the energy level of some of the electrons and holes. At a higher energy level, the electrons and holes are no longer bound in the lattices of the semiconductors. These electrons and holes with kinetic energy will move from the higher-temperature end to the lower-temperature end. Therefore, electric current is generated in the circuit.

The Seebeck coefficient of a material denoted by S, or sometimes by α, is called TE power or thermopower. It is defined as the magnitude of an induced TE voltage ΔV or ΔE in response to a temperature difference across the material ΔT, that is $S = \Delta V/\Delta T = \Delta E/\Delta T$. Evidently, the unit of the Seebeck coefficient is volts per kelvin (V/K) or microvolts per kelvin (μV/K). How strong the thermopower of a material will be depends on the structure of the material and the temperature difference. Metallic materials have very small Seebeck coefficients because of their half-filled bands. Since electrons (negative charges) and holes (positive charges) both contribute to the induced TE voltage, they cancel each other's contribution to the voltage and make it very small. Later, semiconductors were found. These materials are doped with other elements so that there exists an excess amount of electrons or holes. The charge of the excess carriers makes it possible for semiconductors to have large positive or negative values of Seebeck coefficients. The sign of the Seebeck coefficient determines which type of carrier dominates the charge transport in TE materials.

It should be noted that it is very difficult to obtain the absolute value of the Seebeck coefficient or the thermopower of the material of interest (denoted as material A with the Seebeck coefficient of S_A). During Seebeck coefficient measurement, electrodes attached to a voltmeter must be placed onto the material in order to determine the TE voltage.

The temperature gradient then also induces a TE voltage across one leg of the measurement electrodes (with a Seebeck coefficient of S_B). Therefore, the measured thermopower (S_{AB}) includes a contribution from the thermopower of the material of interest and the material of the measurement electrodes. The measured thermopower is then a contribution from both and can be expressed by Eq. 1.5. By using superconductor electrodes, the absolute Seebeck coefficient can be directly measured. In superconductors, the charged carriers produce no entropy, which means they contribute no TE voltage. Therefore, superconductive materials have zero thermopower.

$$S = S_{AB} = S_A - S_B = \frac{\Delta V_A}{\Delta T} - \frac{\Delta V_B}{\Delta T}. \quad (1.5)$$

1.1.2 Peltier Effect

The reversed energy conversion mechanism from the Seebeck effect is called the Peltier effect, which was discovered in 1834 by the French physicist Jean-Charles-Athanase Peltier. In the Peltier effect, an TE material converts electrical energy into thermal energy.

Conceptially, when an electric current I passes through TE materials, a temperature difference is established, that is $I \Rightarrow \Delta T$. One end of the TE materials is under active cooling while heat is rejected to the other end. Due to the electric potential difference generated by an external power, the charge carriers in the semiconductors as shown in Fig. 1.3 migrate from one end to another.

The movement of the electrons in the n-type semiconductor results in a transfer of the internal energy from the top part to the bottom part, and the upper end cools. The movement of the holes in the p-type semiconductor results in the same effect. Therefore, Fig. 1.3 represents a TE pump module. The heat transfer from the cooling end to the hot end, Q, is proportional to the carrier current passing through the circuit, that is

$$Q = \Pi_{AB} I = \left(\Pi_{AB} - \Pi_{AB}\right) I, \quad (1.6)$$

where Π_{AB} is the Peltier coefficient of the semiconductor/copper wire junction, Π_A is the Peltier coefficient of the semiconductor, and Π_B is the Peltier coefficient of the copper wire.

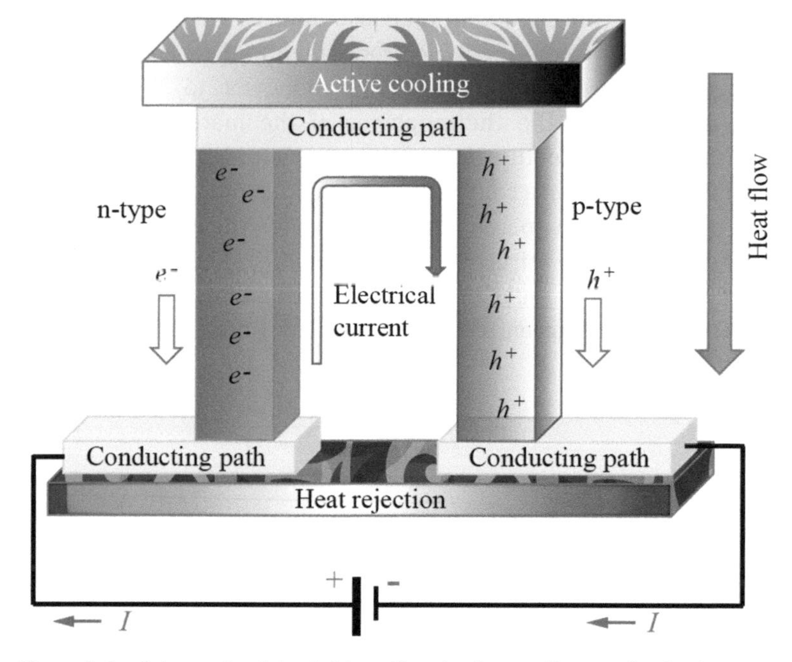

Figure 1.3 Schematic of the Peltier effect (active cooling mechanism).

1.1.3 Thomson Effect

When an electric current I moves in a material, heat generation can be expressed as

$$Q = \pm \mu I \frac{dT}{dx} + IR^2, \quad (1.7)$$

where μ is the Thomson coefficient, R is the electrical resistance of the TE material of interest, and $\frac{dT}{dx}$ is the temperature gradient along the TE material.

Among the three TE coefficients, S, Π, and μ, the Thomson coefficient, μ, is unique because it is the only TE coefficient directly measurable for individual materials. The term IR^2 in Eq. 1.7 is due to joule heating. It is irreversible. The other term in Eq. 1.7 is the contribution of the Thomson effect, which is called Thomson heat. It changes sign when I changes the direction.

It is evident that the Thomson effect describes the heating or cooling of a current carrying conductor with a temperature gradient. William Thomson (Lord Kelvin) predicted the Thomson effect and experimentally confirmed the effect in 1851. For any current carrying material (except for a superconductor), if there exists a temperature difference between two points, depending on the nature of the material, it either absorbs or emits heat.

Some materials have the positive Thomson effect. For example, in copper and zinc, which have a hotter end at a higher potential and a cooler end at a lower potential, when current moves from the hotter end to the colder end, it is moving from a high to a low potential. Heat is generated in the materials. In contrast to this, some materials, such as iron, nickel, and cobalt, show the negative Thomson effect. In these metals, which have a cooler end at a higher potential and a hotter end at a lower potential, when current moves from the hotter end to the colder end, it is moving from a low to a high potential. Heat is absorbed by the materials. Lead is considered to have an almost zero Thomson effect because the TE coefficient of lead is very small. In addition, the TE coefficient of all known superconductors is zero.

The Peltier and Seebeck coefficients can only be determined for pairs of materials, that is, in the form of Π_{AB} and S_{AB}). If no superconductor is used, there is no direct experimental method to measure the absolute Seebeck coefficient (thermopower), S, or the absolute Peltier coefficient, Π, for an individual material. However, the Seebeck effect is actually a combination of the Peltier and Thomson effects. In 1854 Thomson found two equations, called the Thomson relations, also known as the Kelvin relations, relating the three TE coefficients. These relations are expressed as

$$\mu = T\frac{dS}{dT} \tag{1.8a}$$

and

$$\Pi = TS. \tag{1.8b}$$

The Thomson coefficient, μ, is unique. Its absolute value can be determined directly by experiments.

If the Thomson coefficient of a material is measured over a temperature range, one can then integrate the Thomson coefficient over the temperature range using the first Kelvin relation shown by

Eq. 1.8a to calculate the absolute value of the Seebeck coefficient. Then Eq. 1.8b can be used to determine the absolute value of the Peltier coefficient, Π.

1.2 Thermoelectric Energy Conversion Efficiency

Heat is a form of energy. According to the first law of thermodynamics, energy is conserved. Thus, for a heat engine doing work by absorbing heat from a high-temperature source and rejecting heat at a low-temperature source, as shown in Fig. 1.4, the following relation holds:

$$Q_H = W + Q_L, \tag{1.9}$$

where W is the work done by the heat engine. The efficiency of the energy conversion is defined by

$$\eta = \frac{W}{Q_H} = \frac{Q_H - Q_L}{Q_H}. \tag{1.10}$$

The Carnot limit yields

$$\eta \le \eta_c = \frac{T_H - T_L}{T_H}, \tag{1.11}$$

where η_c is the energy conversion efficiency for a system following the Carnot cycle.

T_H : Temperature of the heat source
T_L : Temperature of the heat sink
Q_H : Heat rejected by the source
Q_L : Heat received by the sink
W : Work done

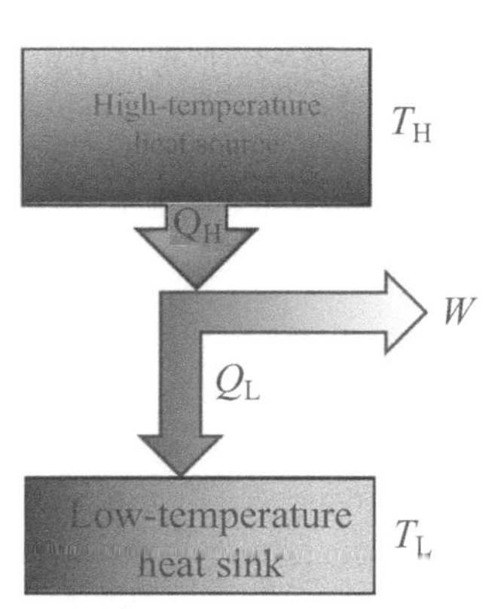

Figure 1.4 Energy conversion in a heat engine.

Therefore, the upper limit of energy conversion efficiency is given by

$$\eta \leq 1 - \frac{T_\mathrm{L}}{T_\mathrm{H}} = \frac{\Delta T}{T_\mathrm{H}}. \tag{1.12}$$

From the above analysis, it can be seen that heat-to-electricity conversion undergoes the Carnot limit. If T_H < 100°C, a very low-temperature heat sink is needed. This adds to cost. In large-scale energy conversion power plants, the Rankine cycle is applied and the overall plant efficiency is in the range from 34% to 42%. But such power plants are difficult to scale down. Small-scale thermal-energy-to-electric-energy conversion systems with power generation capacity less than 10 kW operate following the Stirling cycle. Such engines have a higher energy conversion efficiency, of up to 50%. Some shortcomings are associated with these engines. They are heavy in weight and vibrate in operation. Besides, they are not scalable. Thermoelectric generators (TEGs) use solid-state semiconductors. The efficiency is in the range of 10%–20%. As one of the advantages mentioned before, TE energy converters are scalable. The performance of TE materials can be evaluated by a material parameter called the figure of merit *Z*, which is expressed as [1–3]:

$$Z = \frac{\sigma S^2}{\kappa}, \tag{1.13}$$

where *S* is the Seebeck coefficient, κ is the thermal conductivity, and σ is the electrical conductivity.

It should be noted that the thermal conductivity includes contributions from both electron transfer and phonon transfer, that is $\kappa = \kappa_\mathrm{e} + \kappa_\mathrm{p}$, where κ_e is the heat conductivity of electrons and κ_p the heat conductivity of phonons.

The energy conversion efficiency of TE materials can be correlated to the dimensionless figure of merit *ZT* [4], that is

$$ZT = \frac{\sigma S^2}{\kappa} T, \tag{1.14}$$

where *T* is the absolute temperature and *Z* is the power factor. Typically, *ZT* = 1 corresponds to energy conversion efficiency of about 10%. The relation between the energy conversion efficiency and the absolute figure of merit is [5]

$$\eta_{\mathrm{TE}} = \eta_{\mathrm{c}} \frac{\sqrt{1+ZT}-1}{\sqrt{1+ZT}+\frac{T_{\mathrm{L}}}{T_{\mathrm{H}}}}, \tag{1.15}$$

where η_{c} is the Carnot efficiency.

Obviously, greater values of ZT indicate a higher TE efficiency. Therefore, ZT is a very convenient figure for comparing the potential efficiency of devices using different TE materials. A value of $ZT = 1$ is considered good, and values in the range of 3–4 are considered to be essential for thermoelectrics to compete with mechanical generation and refrigeration in efficiency. To date, the best reported ZT values have been in the 2–3 range [6]. Much research in TE materials has focused on increasing the Seebeck coefficient and reducing the thermal conductivity, especially by manipulating the nanostructure of the materials [7, 8].

1.3 Advantages of Thermoelectric Energy Conversion

There are many advantages of energy conversion systems based on thermoelectricity. For example, moving parts are not necessary for generating electric power from heat or active cooling by electricity; thus the maintenance needed for these energy conversion systems is substantially less. TE systems have long lifetimes. They can operate for more than 200,000 hours. In addition, TE systems for cooling contain no chlorofluorocarbons, as used in conventional freezers. This is good for environment protection. In TE heating or cooling systems, the temperature can be controlled to within a fraction of a degree. Therefore, TE systems are considered to function well, especially in environments that are too severe, too sensitive, or too small for conventional heating or refrigeration. Typical applications of TE systems include environmentally friendly air conditioning, spot cooling of electronic chips, superconductors, etc. They are also considered to make thermal suits for firefighters, soldiers, etc.

Energy harvesting based on thermoelectricity is very attractive. Waste heat recovery from automobiles, road trucks, utilities, chemical plants, etc., using TE materials is considered. For example, German automobile makers Volkswagen and BMW have developed TEGs that recover waste heat from combustion engines. Volkswagen claims a 600 W output from the TEG under highway driving condition. The TEG-produced electricity meets around 30% of the car's electrical requirements, resulting in a reduced mechanical load (alternator) and a reduction in fuel consumption of more than 5%. BMW and DLR (German Aerospace) have also developed an exhaust-powered TEG that achieves a maximum of 200 W and has been used successfully for road use of more than 12,000 km. Geothermal power generation and remote power generation in space become possible due to thermoelectricity. Space probes to the outer solar system make use of the effect in radioisotope TEGs for electrical power. In addition, solar heat harvesting and converting heat from nuclear reaction directly to electricity can be realized using high-performance TE devices. Miniature TE power generators that directly convert heat into electricity can be made with an array of vertically integrated film thermocouples on thin, high-thermal-conductivity substrates.

References

1. Shoko, E. (2014). Novel K rattling: a new route to thermoelectric materials? *J. Appl. Phys.*, **115**(3), pp. 33703–33708.
2. Talapin, D. V. (2014). Thermoelectric tin selenide: the beauty of simplicity, *Angew. Chem. Int. Ed.*, **53**(35), pp. 9126–9127.
3. Reddy, P. (2014). Electrostatic control of thermoelectricity in molecular junctions, *Nat. Nanotechnol.*, **9**(11), pp. 881–885.
4. Xiao, F., Hangarter, C., Yoo, B., Rheem, Y., Lee, K. H., Myung, N. V. (2008). Recent progress in electrodeposition of thermoelectric thin films and nanostructures, *Electrochim. Acta*, **53**(28), pp. 8103–8117.
5. Tritt, T. M., Bäottner, H., Chen, L. (2008). Thermoelectrics: direct solar thermal energy conversion, *MRS Bull.*, **33**(4), pp. 366–368.
6. Venkatasubramanian, R., Siivola, E., Colpitts, T., O′Quinn, B. (2001). Thin-film thermoelectric devices with high room-temperature figures of merit, *Nature*, **413**(6856), pp. 597–602.
7. Majumdar, A. (2004). Thermoelectricity in semiconductor nanostructures, *Science*, **303**(5659), pp. 777–778.
8. Snyder, G. J., Toberer, E. S. (2008). Complex thermoelectric materials, *Nat. Mater.*, **7**(2), pp. 105–114.

Chapter 2

Micronanoscale Device Concepts and Manufacturing Techniques

In this chapter, micro- and nanoscale device concepts are introduced. Various micronanofabrication and manufacturing techniques are reviewed. The organization of this chapter is as follows: The first part deals with device fabrication and manufacturing through chemical approaches, which include the solution method, chemical vapor deposition (CVD), and multiphase (vapor-liquid-solid) growth. In the second part, high-energy-beam-assisted catalytic growth is discussed. Afterward, electrochemical approaches, including anodizing, electrochemical etching, electroplating, and electrocodeposition, are introduced. Device fabrication through templates is presented. The subsequent section is on micronanoscale manipulation via lithographic and scanning probe techniques. Finally, the electrospinning technique for micronanoscale device fabrication and manufacturing is discussed.

2.1 Introduction

Design and fabrication of microelectromechanical systems (MEMS) and nanoelectromechanical systems (NEMS) have caught increasing interest due to the wide applications. In such systems, various sensors play a critical role. Sensors can be viewed as energy

Nanomaterials for Thermoelectric Devices
Yong X. Gan

ISBN 978-981-4774-98-7 (Hardcover), 978-0-429-48872-6 (eBook)
www.panstanford.com

conversion devices that convert signals from mechanical, chemical, optical, thermal, magnetic, and various other energy domains to the electrical energy domain. Actuators work in the reverse way. They convert energy from the electrical energy domain into the mechanical energy domain. Transducers include both sensors and actuators. The concepts of micro- or nanoscale devices come from the unique properties and functions of nanomaterials or structures associated with the devices or systems. For example, fluids containing carbon nanotubes (CNTs) were made for enhanced heat transfer [1–3]. High-density and large-area vertically aligned porous silicon nanowires (Si NWs) were fabricated on the two sides of the silicon substrate using the metal-assisted chemical etching approach. [4] The porous Si NWs show good thermoelectric (TE) properties with the figure of merit (*ZT*) value of 0.493. A special electrorheological (ER) property can be obtained by incorporating nanomaterials. Under an applied electric field, the added nanoscale entities show actuation behaviors, which allow them to be used as propelling components in nanomachines. The synthesis and characterization of $Pb_3O_2Cl_2$ NWs and the ER properties of the NW containing fluids have recently been studied [5]. The ER properties of the nanomaterial suspensions were tested via oscillatory shear experiments. It was found that the viscoelasticity of the nanosystems changes with the intensity of the applied DC electric fields. The actuation behavior for the laden suspensions was observed at low voltages and a very low concentration of the reinforcements (0.0125 wt%).

Various new nanoscale devices using nanofibers are designed. The concept of an "electronic nose" has been proposed by building nanofiber sensor arrays so that high sensitivity can be achieved to discriminate between different chemicals. Sysoev et al. [6] prepared a nanoscopic electronic nose using an array of individual metal oxide (SnO_2, TiO_2, and In_2O_3) nanofiber sensors. The device is capable of discriminating between CO and H_2. Single-walled carbon nanotubes (SWNTs) were used to make a nanoscale device for the label-free detection of DNA hybridization [7]. The effect of SWNT-DNA binding on DNA functionality is revealed by the electrical conductance change between the SWNT and gold contact. Adsorption of trace amounts of chemical vapors at the defect sites in the SWNT produces a large electronic response that dominates the capacitance and conductance sensitivity of the SWNT. It is considered to be because of the increased adsorbate binding energy and charge transfer at defect sites [8]. It is suggested that oxidation defects enhance the sensitivity of a SWNT network sensor to a variety of chemical vapors.

Nanoscale field emission devices have been made with arrays of metal NWs [9], CNTs [10], and $TaSi_2$ nanofibers [11]. Hwang et al. [9] fabricated vacuum tube arrays using anodic aluminum oxide (AAO) nanotemplates. Ni NWs were deposited electrochemically within the pores of the AAO and were used as field emitters. A titanium thin film was evaporated on the templates to seal the pores. The Ti film also served as the anode of the field emission device. Since the distances between the tips of Ni NWs and the anode are much smaller than those of conventional designs, a much lower turn-on voltage is needed to trigger electron emission. In the work performed by Sohn et al. [10], well-aligned CNT field emitter arrays were prepared for potential electron emission applications (such as cold-cathode flat-panel displays) and vacuum microelectronic devices (such as microwave power amplifier tubes). The well-aligned CNT arrays by thermal chemical vapor deposition (CVD) at temperatures below 800°C on Fe nanoparticles were deposited by a pulsed laser on a porous Si substrate. The field emitter arrays are vertically well-aligned CNTs on the Si-wafer substrate. The cathode of the field emitter is an array of CNTs. The cathode and anode were separated by a polyvinyl film with the thickness of about 60 μm. The field emission behavior of the device was tested at room temperature in a vacuum chamber below 10^{-6} Torr. High field emission current densities can be obtained at relatively low electric fields owing to the good adhesion of the CNTs to the Si substrate through Fe nanoparticles. It was found that the CNT field emitter arrays emitted a current of 1 mA/cm^2 at an electric field of 2 V/μm. An emission current density as high as 80 mA cm^{-2} was obtained at 3 V/μm.

Nanoscale field emission devices not only have the advantage of very low turn-on electric fields but also demonstrate the ability to hold at very high current density. For example, $TaSi_2$ NW emitters have a remarkably high failure current density, on the order of 10^{-8} A/cm^2, which promises future applications in nanoelectronics and nano-optoelectronics. Chueh et al. [11] introduced the synthesis of $TaSi_2$ NWs on a Si substrate by annealing $NiSi_2$ films at 950°C in tantalum (Ta) vapor. The obtained NWs are as long as 13 μm. The metallic $TaSi_2$ NWs exhibit excellent electrical properties through field emission measurements. It is found that the turn-on electric field is as low as 4 V μm^{-1}, while the failure current density is as high as 3×10^{-8} A/cm^2.

2.2 Chemical Approaches

Chemical approaches have found wide applications in synthesizing semiconductors, metals, oxides, and various micronanoscale devices. A very common example is the solution-based crystal growth. Through chemical approaches, the structures and performances of many nanoscale materials can be tailored. Chemical approaches allow the study of size-dependent phenomena at the nanoscale and thus provide the potential for exploring new types of devices with precise functions. In the following discussion, three chemical fabrication methods—the solution method, CVD, and vapor-liquid-solid (VLS) growth—will be presented.

2.2.1 Solution Method: Liquid-Phase Deposition

The solution method, or liquid-phase deposition, is commonly used for the synthesis of nanoparticles, nanotubes (NTs), and NWs. Zhang et al. [12] synthesized an yttrium-aluminum-garnet (YAG) nanopowder with aluminum and yttrium nitrates as the starting materials and an ethanol-water mixture as the solvent. The powder is single-phased YAG with an average grain size of 80 nm. The grain size distribution is in a relatively narrow range. The solution method has also been used to prepare oxide sheets with nanoholes, for example, a titanium oxide sheet containing a nanohole array [13]. The titanium oxide with the nanohole array was fabricated through equilibrium reactions in fluorocomplex solutions. It is found that the crystal structure of titanium oxide with the nanohole array is amorphous in an as-deposited state. After heat treatment, it becomes anatase. Spatially organized nanodots were fabricated by exploiting the solution method under controlled dewetting, ripening, and crystallization conditions [14].

The size of the nanodot features can be as small as 100 nm. Functional nanodots and nanorods with different sizes and shapes have recently been fabricated through a simple solution route, as shown by Cao et al. [15]. They synthesized single-crystalline ZnO nanorods with the smallest diameter, of about 5 nm, at ambient temperatures in ethanol without using additional surfactants.

Recently, there has been considerable interest in the fabrication of 1D coaxial layered NTs because of the remarkable properties that are different from those of single-layered NTs. For semiconductors comprising a core and a shell, improved performance can be obtained. A core-shell motif has permitted enhanced photoluminescence, improved stability against photochemical oxidation, and engineered band structures. Coaxial layered NTs are expected to have great potential applications for photoelectronic nanoscale devices. Thus far, different forms of multilayered coaxial nanostructures have been produced by various methods, such as self-assembling [16], layer-by-layer deposition [17], and atomic layer deposition [18]. Carny et al. [16] designed trilayered (metal-insulator-metal) coaxial nanocables that may have unique and useful electromagnetic properties for applications in micro- and nanoscale systems. They fabricated these coaxial NTs via site-specific metal reduction. To generate the chemical reaction sites, gold nanoparticles were bounded to the surface of peptide NTs through a common molecular recognition element that was included in various linker peptides.

Multilayered NTs can be synthesized by a liquid-phase-deposition approach. Figure 2.1 shows the schematic of the liquid-phase deposition and the scanning electron microscopic (SEM) images of the prepared CdS-TiO_2 coaxial NTs. Liquid-phase deposition represents a simple fabrication method. Nanoporous templates may be used as the forming molds for the deposition. For example, this method was used for the preparation of CdS-TiO_2 hybrid coaxial nanocables within porous AAO templates [19]. The thickness of the TiO_2 NTs could be controlled precisely by adjusting the reaction conditions. A polycrystalline CdS layer was deposited onto the titanium oxide NTs during another chemical deposition process so that coaxial CdS-TiO_2 nanocables were formed. Core-sheath nanocables have potential applications in solar energy cells, semiconductor photocatalysis, water purification reactors, and electrochromic devices [20]. The fabrication procedures for the preparation of CdS-TiO_2 hybrid coaxial nanocables include three steps. The first step is the preparation of the TiO_2 NT array through deposition from an aqueous solution mixture of ammonium hexafluorotitanate and boric acid into the nanopores of AAO membranes. The second step is the synthesis of hybrid CdS-TiO_2 NTs. CdS-TiO_2 NTs are synthesized with TiO_2 NTs as the templates via a solution reaction approach by injecting reactants into the pores of TiO_2-AAO templates. In the third step, the CdS-TiO_2 coaxial NTs are harvested by chemically dissolving the AAO templates.

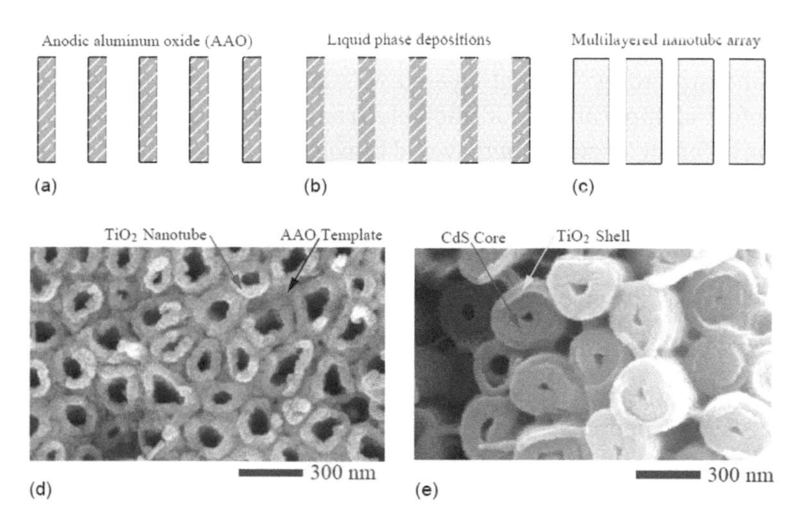

Figure 2.1 Liquid-phase deposition of multilayered nanotubes: (a) schematic of an anodic aluminum oxide (AAO) nanoporous template, (b) multistep liquid-phase depositions, (c) multilayered nanotube array after the removal of AAO, (d) scanning electron microscopic (SEM) image showing the liquid-phase-deposited TiO_2 nanotube in AAO [19], (e) and a SEM image showing the CdS deposited at the inner wall of the TiO_2 nanotube to form a core-shell structure. (d, e) Reprinted from Ref. [19], Copyright (2005), with permission from Elsevier.

2.2.2 Chemical Vapor Deposition

CVD is one of the most useful micro- and nanoscale fabrication techniques. During CVD, heterogeneous reactions may take place not only on a substrate but also in the gaseous phase. As shown in Fig. 2.2a, CVD includes several processes. First, reactants and inert dilute gases flow into the chamber. Then, gaseous species move to the substrate and reactants adsorb onto the substrate. Figure 2.2b is a photograph showing the setup of the CVD equipment.

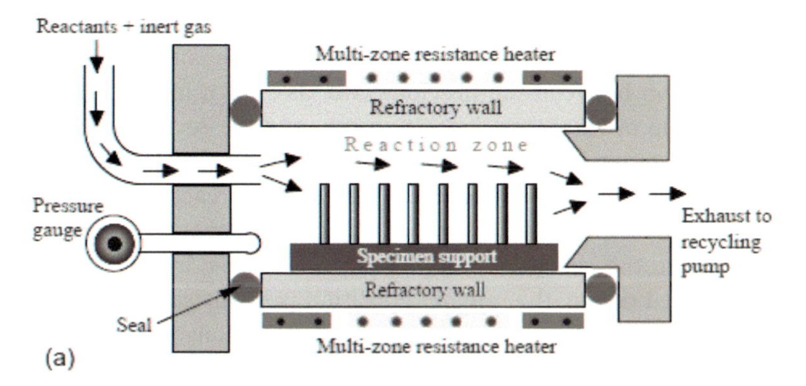

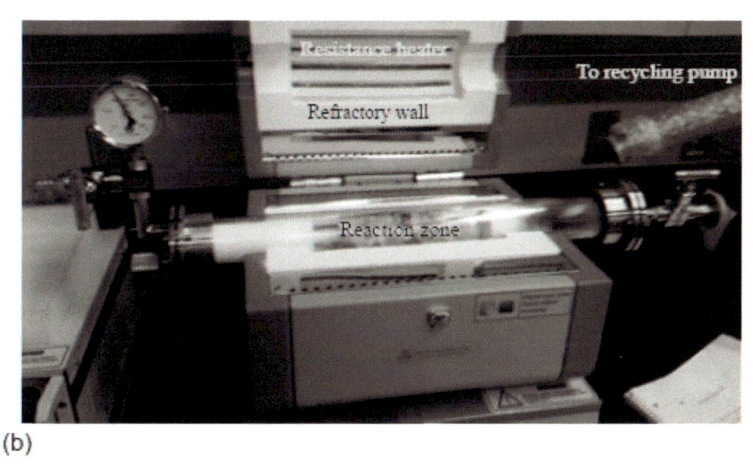

Figure 2.2 Chemical vapor deposition setup: (a) schematic of the process and (b) instrument setup.

The adsorbed atoms migrate and undergo chemical reactions. The gaseous by-products of the reactions are desorbed from the surface of the specimens and removed from the reaction chamber. The exhaust is subsequently trapped in a recycle tank through a vacuum pump. To precisely control the temperature at different locations in the chamber, multizone resistance heaters are used.

The deposition rate in a CVD process is temperature dependent. In the high-temperature range, mass transfer controls the whole process of deposition. Nevertheless, in the low-temperature range, the chemical reaction determines the rate of deposition. Transition between the two mechanisms is also found. Thus, the CVD kinetics can be divided into three regimes. They are the mass transfer–limited regime, the reaction rate–limited regime, and the regime of transition between the previous two, as shown in Fig. 2.3 (the relationship of logarithm deposition rate and the reciprocal of the reaction temperature, $1/T$, with the dimension of K^{-1}). It is noted that in the mass transfer–limited regime, if CVD reactants arrive at the surface of specimens uniformly, the rate of deposition is not so sensitive to the change of temperatures. However, in the reaction rate–limited region, the rate of deposition is highly sensitive to the temperatures. It is very important to keep a uniform temperature distribution in the reaction zone, as shown in Fig. 2.2a, to ensure uniform deposition. Understanding such kinetics will help to control the dimension, morphology, and structure of nanoscale products fabricated from CVD.

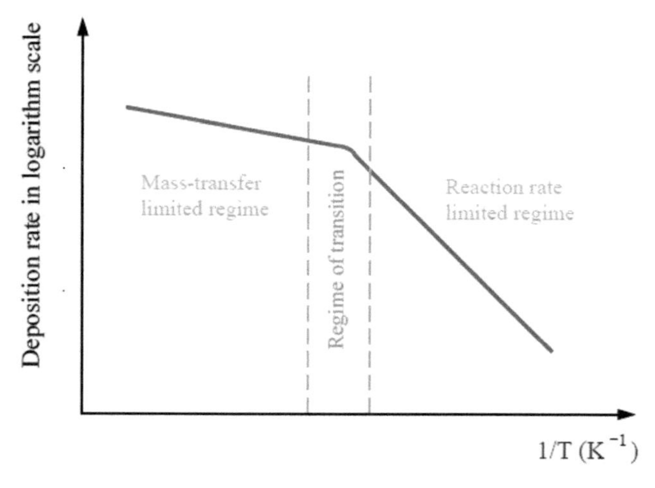

Figure 2.3 Schematic showing the three-stage kinetics of chemical vapor deposition (CVD).

There are different types of CVD processes [21, 22]. One is the so-called atmospheric pressure chemical vapor deposition (APCVD). APCVD is a high-pressure mass transport–limited process. It is susceptible to gas phase reactions. There may be problems such as the formation of particulate by-products and lack of uniformity of coatings. Therefore, the morphology of nanoscale features fabricated by APCVD is very difficult to control. The second type of CVD is the low-pressure chemical vapor deposition (LPCVD). In LPCVD, the pressure level is controlled to less than 1 Torr. It is a reaction rate–limited process. The temperature for LPCVD is usually higher than that for APCVD. Plasma-enhanced chemical vapor deposition (PECVD) is another CVD technique. It is a surface reaction rate–limited process. During PECVD, the energy is supplied by plasma. The major advantages of PECVD are a low processing temperature, a high deposition rate, low porosity, and deposited products of a high quality.

Examples of CVD are shown as follows: the first example is APCVD or LPCVD of SiO_2 nanoscale particles and/or thin films. In a CVD chamber, silane and oxygen can react with each other as shown in Eq. 2.1:

$$SiH_4 + 2O_2 = SiO_2\downarrow + 2H_2O \tag{2.1}$$

Plasma-enhanced or sputtering CVD deposition of SiO_2 can be achieved by the decomposition of tetraethoxysilane (TEOS) [4]. The reaction is shown in Eq. 2.2:

$$Si(OC_2H_5)_4 \rightarrow SiO_2\downarrow + R, \tag{2.2}$$

where R stands for the generated small molecular by-products such as H_2O and CO_2. More examples on CVD fabrication of SiO_2 are given in Ref. [4]. For example, the reaction of dichlorosilane (DCS) and nitrous oxide as shown in Eq. 2.3 is

$$SiCl_2H_2 + 2N_2O = SiO_2\downarrow + 2N_2\uparrow + 2HCl\uparrow \tag{2.3}$$

Silicon nitride, Si_xN_y, as a very useful passivation coating, is extensively used in micro- and nanofabrication [22]. The standard chemical formula for silicon nitride is Si_3N_4. Silicon nitride is inert to almost all kinds of wet etchants except for boiling phosphoric acid. More often than not, patterning silicon nitride using plasma etching is necessary. Because of the special properties of silicon nitrides, they are used as capacitor dielectrics, structural materials, antireflective coatings, masks for etching, and selective oxidization of silicon wafers. Depending on the different starting materials to be used, Si_3N_4 may be obtained by either CVD or LPCVD through the reactions shown in Eqs. 2.4a and 2.4b.

$$3SiH_4 + 4NH_3 = Si_3N_4\downarrow + 12H_2\uparrow \tag{2.4a}$$

$$3SiCl_2H_2 + 4NH_3 = Si_3N_4\downarrow + 6H_2\uparrow + 6HCl\uparrow \tag{2.4b}$$

LPCVD is suitable for fabricating polycrystalline Si. By pyrolyzing nitrogen-diluted silane in a low-pressure reactor in the temperature range of 600°C–650°C, the following reaction holds:

$$SiH_4 = Si\downarrow + 2H_2\uparrow \quad (2.5)$$

Temperature control is very important in the LPCVD process. It is found that at lower temperatures, the reaction of Eq. 2.5 is so slow that the deposition is not practical. If the temperature is too high, the reaction occurs in the gas phase as well as on specimens or substrates. The morphology and properties of the LPCVD-polycrystalline Si are influenced by the temperature, dopants, and post–heat treatment conditions. For example, amorphous films form at 605°C, while columnar structures form at 630°C. The column-like nanostructures were also found by Goerigk and Williamson [23] in hydrogenated amorphous polycrystalline silicon germanium alloys. The hydrogenated amorphous silicon germanium alloys with the composition of α-$Si_{1-x}Ge_x$: H ($0.5 < x < 0.7$) were prepared by different PECVD techniques. The results from anomalous small-angle X-ray scattering (ASAXS) analysis show inhomogeneous distributions of Ge with correlation lengths of 1.0–1.4 nm. Enhanced ion bombardment during growth and hydrogen dilution of the plasma can reduce the nonuniformity of Ge distributions. The more homogeneous materials help to improve the photovoltaic and optoelectronic performance of the nanodevices made from polycrystalline Si.

Another important material, SiC, used in NEMS and MEMS may be synthesized through various approaches of CVD. SiC is extremely stiff and hard; it has very good resistance to chemical attacks by KOH or hydrofluoric acid (HF). Despite the high chemical resistance of SiC, micro- and nanoscale features of SiC can be easily patterned using plasma-enhanced (i.e., SF_6 plasma) techniques. SiC can also be fabricated through PECVD or grow in situ. The reaction is shown in Eq. 2.6:

$$SiH_4 + CH_4 = SiC\downarrow + 4H_2\uparrow \quad (2.6)$$

The resulting SiC is in amorphous form. In some cases, the hydrogenated silicon carbide (α-SiC:H) coexists with SiC in the products.

CVD has been also used to manufacture CNTs in large quantities at a reasonable cost using metal catalysts and carbon monoxide as the carbon source [24]. The structure of CNTs can be adjusted through the selection of the catalysts and control of the thermal conditions used in production [25]. Figure 2.4 is given to illustrate the CVD mechanisms for CNT fabrication. The metal catalysts such as Ni and Fe nanoparticles are replenished in the reaction chamber. The chamber is maintained in the temperature range of 750°C–900°C. Carbon monoxide is introduced through a compressive pump. The pressure in the reaction chamber is kept at 10 atm. The chemical reaction for CNT deposition can be expressed as Eq. 2.7:

$$2CO = C\downarrow + CO_2\uparrow \tag{2.7}$$

On top of the reaction chamber, as shown in Fig. 2.4, there is a product collecting unit. The CVD-fabricated CNTs are harvested through an outlet connected to the product collecting unit. The other product, carbon dioxide, is trapped in a storage tank. Furthermore, the exhausted CO can be recycled using a recycling pump.

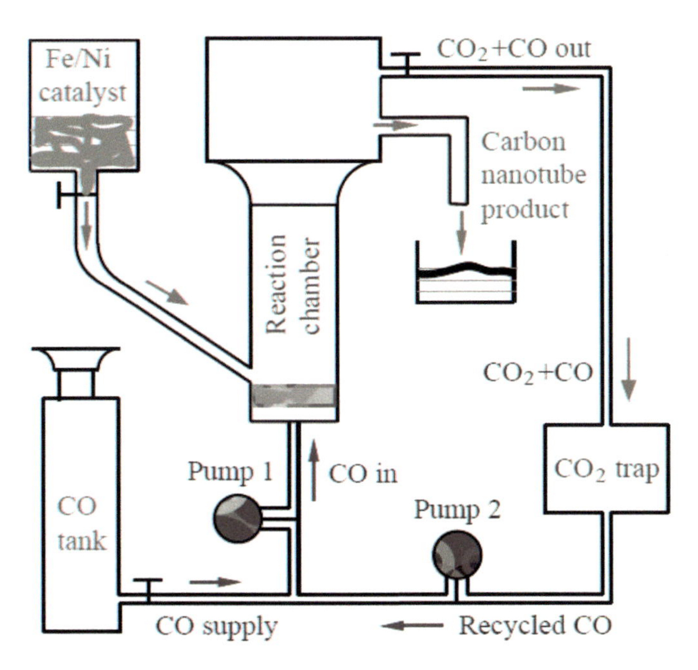

Figure 2.4 CVD fabrication of CNTs.

There is considerable progress in the design and development of new CVD facilities. Recently, an integrated PECVD nanofabrication facility has been set up for assembling quantum structures and fabricating functional nanomaterials [26]. The key component of the integrated plasma-aided nanofabrication facility (IPANF) is a special configured multiple-target radio frequency (RF) magnetron sputtering plasma source. Two configurations of the plasma source, an external flat spiral coil configuration and an internal antenna configuration comprising two orthogonal RF current sheets, were designed. Unidirectional RF currents can be obtained from the internal antenna, which generates uniform plasma over large areas and volumes. The IPANF was used for various nanoscale fabrications, such as the low-temperature PECVD of vertically aligned single-crystalline carbon nanotips; the growth of ultrahigh-aspect-ratio semiconductor NWs; the assembly of Si, SiC, and $Al_{1-x}In_xN$ quantum dots with optoelectronic functions; and the plasma-based synthesis of bioactive hydroxyapatite for orthopedic implants.

Exploration of new CVD technology is still underway. Electron cyclotron resonance chemical vapor deposition (ECR-CVD), as an innovative approach, has been used for the nanofabrication of GeSbTeSn phase-change alloy-ended CNTs [27]. Cobalt-assisted CNTs were synthesized first with H_2 and CH_4 as the gaseous reactants to provide a carbon source. Then, the as-grown CNTs were treated in a H plasma atmosphere to remove the carbon layers on the Co catalyst. The catalyst was subsequently removed from the tips of the CNTs in the HNO_3 solution. The open-ended CNTs with bowl-like tips were coated with a layer of the phase change alloy. Structure analysis reveals that the composition of the alloy on the tips of CNTs changes from Te rich to Ge rich after heat treatment. Such CNTs with the phase change alloy have potential applications as nanoresolution storage media. Microwave plasma chemical vapor deposition (MP-CVD) was used for the preparation of highly oriented carbon nanocones [28]. The nanostructures of Co-assisted carbon nanocones (CNCs) were manipulated by adjusting the ratios of the source gases (CH_4 and H_2) and the substrate bias. The formation of the cone-shaped nanostructures is the result of competition between the ion bombardment and the lateral growth of the nanostructures in the plasma. Since the ion bombardment is enhanced by the negative substrate bias, a higher substrate negative bias and a lower concentration of carbon species in the plasma are favorable conditions to grow highly oriented CNCs owing to a greater ion bombardment energy and a lower lateral growth rate of the nanostructures. The CNCs were characterized as field emitters and showed excellent performance. It is found that the CNCs synthesized under the applied bias of –300 V provide a high field emission current density of 173 mA/cm^2 at 10 V/μm.

2.2.3 Vapor-Liquid-Solid Growth

Vapor-liquid-solid (VLS) growth of nanostructures involves multiphase catalyst–assisted heterogeneous chemical reactions. The reactants typically come from the gas phase. The catalysts used are in the liquid state as droplets set on the solid substrates, as shown in Fig. 2.5a. Figure 2.5b shows the saturation of the reactants, A and B, in the liquid phase. The reaction of A and B generates product C, and the heterogeneous nucleation of C on the substrate is shown in Fig. 2.5c. Because of selective wetting of the reactants to the catalyst, the reaction continues in the liquid-solid interface region; thus, directional growth of the product into 1D nanostructures such as fibers and wires occurs as illustrated in Fig. 2.5d. Wang et al. [29] investigated Au-catalyzed VLS growth of phosphorus-doped Si NWs. Silane (SiH_4) gas was the source of Si, and phosphine (PH_3) was used as the n-type dopant source. The as-grown Si NWs even have a single-crystal structure. The VLS-fabricated Si NWs containing electrically active phosphorus are suitable for electronic and optoelectronic device applications.

Shape and orientation control of germanium NWs during VLS growth was performed by Adhikari et al. [30]. The preferential growth of the NWs along the <111> crystallographic orientations was achieved on different substrates at a temperature of 350°C. Vertically aligned NWs were obtained on Ge (111) wafers. In addition to <111> growth, <110> growth was observed on Ge (001) and Ge (110) substrates. The structure of nanostructures fabricated through the VLS approach is influenced by the reaction temperature. For example, core-shell NWs with the composition of GaAs-$Ga_xIn_{1-x}P$ ($0.34 < x < 0.69$) were synthesized by changing the reaction temperatures [31]. The NW core was grown under the Au-catalyzed assistance at a low temperature (450°C), while the shell was added by growth at a higher temperature (600°C). It is evident that at lower temperatures, the side facet growth is inhibited. At higher temperatures, the kinetic hindrance of the side facet growth is hindered.

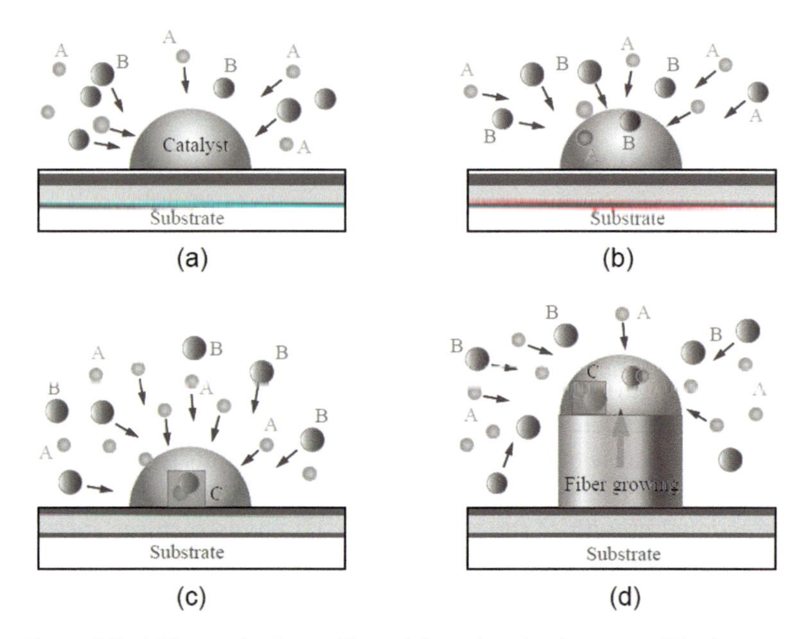

Figure 2.5 VLS growth of nanofibers: (a) catalyst droplet on a solid substrate, (b) dissolving of reactants "A" and "B" from gas phase into liquid phase, (c) catalyst-assisted reaction of "A" and "B" in the liquid phase and heterogeneous nucleation of the product "C" on the solid substrate, and (d) directional growth of the nanofiber.

VLS growth mechanisms involved with the so-called self-catalysis schemes were studied [32]. The growth of indium nitride (InN) NWs in a dissociated ammonia environment provides insight into the nucleation and growth mechanisms. The nucleation of the InN crystal occurs first on the substrate, which is very similar to that shown in Fig. 2.5c.

Indium droplets stayed on top of the InN crystals because indium preferentially wets the InN crystals. The continued reaction through liquid-phase epitaxy ensures basal growth, resulting in the formation of 1D nanostructures, for example, InN NWs, which may be schematically shown as in Fig. 2.5d. "Tree-like" bifurcated morphologies on a variety of substrates were also obtained via a heteroepitaxial growth process. However, the direct nitridation of indium droplets on a single-crystal substrate using dissociated ammonia results in the spontaneous nucleation and basal growth of the InN NWs, which is typical in VLS processes.

2.3 Laser-Assisted Catalytic Growth

Laser-assisted catalytic growth [21, 22, 33] is similar to the VLS technique in terms of involving catalyst-assisted, multiphase chemical reaction and deposition processes. However, the reactants for laser-assisted catalytic growth are not only from reactive gases but also from the bombardments of laser beams. As schematically shown in Fig. 2.6a, the facilities for laser-assisted catalytic growth include a laser source and all the other components for liquid-vapor-solid (LVS) fabrication. The laser target material may be a pure metal or an alloy. Other kinds of materials, such as polymers and ceramics, can also be used as target materials depending on the nature of the nanostructures to be fabricated. In some cases, more than one target material may be used.

Figure 2.6b shows the progressive growth of 1D nanostructures via laser-assisted catalytic reactions. Five growth stages are shown in this drawing. In the first stage, the catalyst is set on the substrate. Catalysts such as Fe, Co, Ni, In, and Au may be prepatterned on the substrate, for example, a Si wafer, via lithographic fabrication techniques, as will be discussed later. These catalysts contract and become spherical in shape and partially wet the substrate at elevated temperatures in the furnace, as shown in Fig. 2.6a. The third stage, as shown in Fig. 2.6b, is the LVS growth of the substance M when the laser source is in an off state. In the fourth stage, the laser source is on and high, intensive beams hit the solid target. The growth of another substance, named N, dominates the deposition processes. In the fifth stage, alternative growth of the materials M and N proceeds because of the intermittent bombardment of the target by the programmed laser pulses. Therefore, 1D nanostructures such as NWs, nanofibers, nanocones, and core-shell coaxial NTs may be fabricated through laser- and catalyst-assisted processes.

One of the advantages of laser-assisted catalytic growth is that it provides the capability to fabricate longitudinally ordered heterostructures. For example, heterojunctions and superlattices

(SLs) of Si-SiGe were fabricated using the hybrid pulsed laser ablation–vapor-liquid-solid chemical vapor deposition (PLA–VLS CVD) process [33]. The ablation of the target material by programmed laser pulses provides a pulsed vapor source. This allows the NW growth in a block-by-block fashion of "M–N–M–N–M–N," as shown in Fig. 2.6b, where M stands for Si and N represents SiGe. The NW in a single-crystalline form has a SL structure and shows a well-defined compositional profile along the wire axis. Semiconductor NWs with longitudinal ordered heterostructures are essential for many potential applications of semiconductor NWs in nanoscale optoelectronics. The unique properties of the heterostructured 1D nanostructures also allow them to be used as nanoscale light emitter and TE components. Therefore, laser-assisted catalytic growth is expected to play an important role in the fabrication of nanoscale materials, structures, devices, and systems.

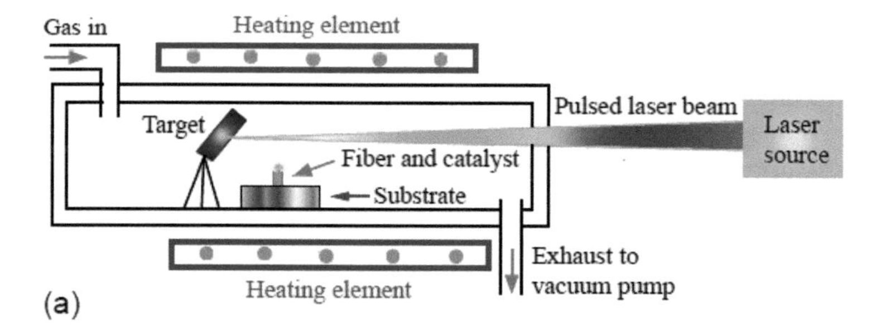

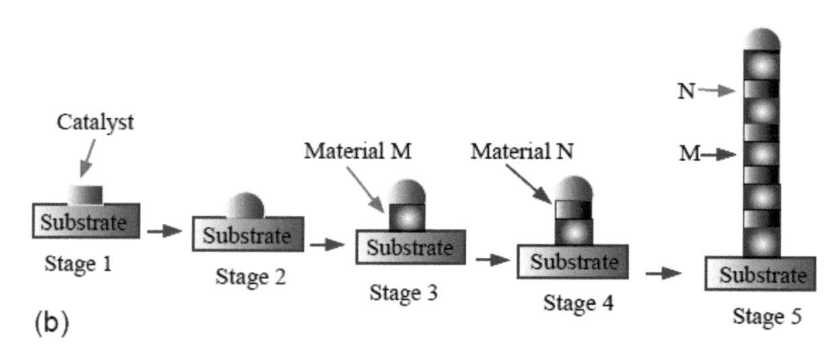

Figure 2.6 Schematic of laser-assisted nanofabrication: (a) setup of the reactor and (b) different stages of laser- and catalyst-assisted growth of 1D nanostructures.

2.4 Electrochemical Approaches

Nanofabrication through electrochemical approaches provides a simple way to obtain nanofibers, NWs, and nanoporous materials and structures. The facility for electrochemical manipulation is very simple, and the electrochemical reactions occur at ambient temperatures. The rate of deposition and dissolving associated with electrochemical approaches can be well controlled. There arc various methods related to electrochemical approaches. Electrochemical oxidization, electrochemical etching, electroplating, and electrocodeposition are some of the examples. A brief review of these methods is given.

2.4.1 Electrochemical Oxidization

Electrochemical reactions on anodes are oxidization processes. Shape- and size-selective electrochemical synthesis is possible through controlled oxidization. Recent studies [34] demonstrated the feasibility of electrochemical fabrication of dispersed silver (I) oxide (Ag_2O) nanoparticles through anodizing of a sacrificial silver wire in a basic aqueous sulfate solution. The Ag_2O particles were released from the surface of the silver electrode during synthesis, producing a trace of colloidal flow. The process of the formation of the Ag_2O nanoparticles may be considered involving three stages. In the first stage, the oxidization of Ag occurs on the surface of the anode following the reaction shown in Eq. 2.8:

$$Ag^{o} - e^{-} = Ag^{+} \quad (2.8)$$

In the second stage, the Ag^+ is associated with OH^- to form a silver hydraulic compound, as shown in Eq. 2.9:

$$Ag^{+} + OH^{-} = AgOH \quad (2.9)$$

AgOH is in a gelation state, existing in the double layer of the anodic region. The last stage is the precipitation of the Ag_2O nanoparticles owing to the dehydration of AgOH, as shown in Eq. 2.10:

$$2AgOH = Ag_2O\downarrow + H_2O \quad (2.10)$$

The composition of the particles was determined by various methods, including selected area electron diffraction, X-ray diffraction

(XRD), and X-ray photoelectron spectroscopy. The shape of Ag_2O crystallites can be controlled by adjusting the potential of the silver wire electrode. Ag_2O particles of different shapes were obtained. There were cubic particles with conclave facets (hopper crystals), which grew within 100 mV of the voltage threshold for particle growth. There were also eightfold symmetric "flower"-shaped particles that were formed on applying more positive voltages. It is assumed that a more positive voltage results in the preferential growth in the <111> crystallographic direction, which is the reason for the formation of flower-shaped particles. The diameter of the flower-shaped particles can be controlled in the range from 250 nm to 1.8 μm. The strategy for synthesizing Ag_2O nanoparticles may be naturally extended for obtaining other metallic oxide nanoparticles, such as ZnO, NiO, and ZrO_2.

Nanodimensional multilayered Ta-Ta_2O_5 films were fabricated by the electrochemical oxidization technique as reported by Mardilovich and Kornilovitch [35]. During the electrochemical nanofabrication, a thin layer of tantalum was first sputtered on a smooth insulating substrate. Then, Ta was partially electrochemically oxidized (anodized) to form a Ta_2O_5 layer. The next layer of Ta was deposited on top of Ta_2O_5, and the process was repeated several times so that multilayered structures were formed. The anodization of Ta has a strong planarization tendency, which leads to smooth interfaces between the Ta layer and the Ta_2O_5 layer. The resulting Ta_2O_5 layers are amorphous and free of pinholes. The thickness of both Ta and Ta_2O_5 is dependent on several critical factors, such as the rate of Ta consumption and the anodization conditions. The Ta layers are as thin as 2.8 nm. Other metals, such as Cr, Ni, Ti, and Al, have properties similar to those of Ta because the oxides of these metals are very dense and smooth. Through electrochemical anodization of these metals in nitric acid or other oxidization acids, controlled growth of oxide films is possible. In addition to fabricating nanoparticles and multilayered films, electrochemical anodization can be used to synthesize nanoporous structures, as will be discussed.

2.4.2 Electrochemical Etching

Electrochemical etching involves anodic reaction processes. This technique has been used to prepare Si nanoporous structures [36].

Recently, electrochemical etching of silicon surfaces was performed through a Nafion-membrane mask to implement pore-directed nanolithography [37]. Sato et al. [38] fabricated a picoliter volume glass tube array through a photoassisted electrochemical etching process. The picoliter volume SiO_2 glass tube array is partially embedded in an n-type Si (100) wafer. The protocol for such nanofabrication can be briefly addressed as follows. A patterned gold-chromium thin film coated on the back surface of the substrate was used as a shade mask to define the locations for the selective formation of the array. The electrochemical etching reaction was controlled by the illumination condition so that the pore array was obtained. To form a glass layer, the inner walls of the pores were wet-thermally oxidized through the reaction in Eq. 2.11 [3, 4]:

$$Si + 2H_2O = SiO_2 + 2H_2\uparrow \qquad (2.11)$$

The bulk Si region can be removed by alkaline etching through the reaction in Eq. 2.12 [5]:

$$Si + 2OH^- + 2H_2O = Si(OH)_2^{2+} + 2H_2\uparrow \qquad (2.12)$$

If the bulk Si is completely removed, the glass tubes are released from the substrate. The remaining parts, owing to the partial removal of the bulk Si, are "glass tubes" aligned on the Si wafer. The depth, the exposed region, and the wall thickness of each glass tube may be controlled by adjusting parameters such as the duration of the Si electrochemical etching, the alkaline etching, and the wet-thermal oxidation, respectively. Such nanoscale channel arrays have potential applications in microreactors.

Electrochemical etching of undoped hydrogenated amorphous silicon layers in hydrofluoric acid (HF) solutions was performed by Gros-Jean et al. [39]. The reaction mechanisms are shown in Eq. 2.13:

$$Si - 4e^- = Si^{4+} \qquad (2.13a)$$

$$2H_2O + 2e^- = 2OH^- + H_2\uparrow \qquad (2.13b)$$

$$Si^{4+} + 4OH^- = Si(OH)_4 \qquad (2.13c)$$

$$Si(OH)_4 = SiO_2\downarrow + 2H_2O \qquad (2.13d)$$

$$SiO_2 + 4HF = SiF_4\uparrow + 2H_2O \qquad (2.13e)$$

Anodic oxidization of Si introduces a Si suboxide into the synthesized structures. Such nanoporous Si structures have unique optical properties. The photoluminescence and X-ray photoelectron spectra of the porous silicon prepared by electrochemical etching have been studied by Dimova-Malinovska et al. [40]. The photoluminescence intensity was found depending on the oxidized valence states of Si. Porous silicon with suboxide states exhibiting more intense photoluminescence was confirmed.

2.4.3 Electroplating

Electroplating employs cathodic reactions to deposit metals or metallic alloys. As a mature technology, electroplating has been extensively used to obtain metallic coatings for decoration, surface protection, and corrosion control. The advantage of electroplating is that metals can be deposited on cathodes in a conformal way, which means that metallic coatings can be deposited uniformly on cathodes with complex shapes. Thus, it is possible to fabricate complex architectures using the electrodeposition technique. Recently, electroplating or electrodeposition has been used for nanofabrication. Patterned self-assembled monolayers (SAMs) on solid surfaces were electrochemically deposited at selected regions. Simple deposition steps may be used to create islands or lines in the patterned SAMs. The strategies for creating more complex structures, including via-like structures and multicomponent lines, were presented, and finite size effects were exploited to produce arrays of single-crystal islands [41].

Fabrication and characterization of copper nanofiber arrays in the pores of polycarbonate (PC) membranes were performed [42]. The electrochemical reactions on the anode when copper is dissolving can be expressed by Eqs. 2.14a and 2.14b. Equation 2.14a is the main reaction. The reactions on the cathode during copper deposition can be expressed by Eqs. 2.14c and 2.14d, with Eq. 2.14c as the main reaction.

$$Cu - 2e^- = Cu^{2+} \tag{2.14a}$$

$$4OH^- - 4e^- = O_2\uparrow + 2H_2O \tag{2.14b}$$

$$Cu^{2+} + 2e^- = Cu \tag{2.14c}$$

$$2H^+ + 2e^- = H_2\uparrow \tag{2.14d}$$

The track-etch membranes of a PC have a nominal pore size of 800 nm. The effects of various parameters during electrodeposition were studied. The wettability of the electrolyte and the PC membranes is critical for the uniform growth of the nanofibers; thus, proper wetting of the membranes prior to electrodeposition was carried out. Controlling of the cathodic current density is important because high-quality nanofibers can only be obtained in a narrow range of current densities owing to the existence of the reaction in Eq. 2.14d. The actual diameter of the nanofibers is slightly larger than the nominal pore size. It is found that the material has a face-centered cubic (FCC) crystalline structure with a high texture coefficient for (200) planes.

2.4.4 Electrocodeposition

Electrocodeposition is a process by which both metals and nanoparticles are deposited on the surface of cathodes to form composite films. Figure 2.7 shows the setup for electrocodeposition of metals and nanoparticles. In Fig. 2.7a, nanoparticle-reinforced composites are deposited on a conductive plate connected to the rotating cathode, while in Fig. 2.7b, nanocomposites are deposited on the stationary cathode. Typically, the content of nanoparticles in the deposited films is in the range of 1%–10% in volume. This process has many advantages, such as the ability to deposit on complex surfaces, the reduction of waste, and the capability to synthesize functionally gradient materials [43]. In addition, electrocodeposition is usually operated at an ambient temperature. The reaction mechanisms of electrocodeposition as well as techniques that promote the dispersion of nanoparticles in metal matrices have been extensively studied [44–51]. Incorporating nanoparticles into nanocomposites is expected to improve the mechanical properties of the materials. Typically, the hardness of nanocomposite thin films increases with the increase of the particle content. In addition, the wear resistance of the nanocomposite thin films can be significantly increased.

Earlier studies have demonstrated that low-porosity metal matrix nanocomposite materials can be prepared with electrocodeposition technology [52–56]. For example, electrocodeposition of nanometer-sized alumina particles in a copper matrix was studied using a concentric cylindrical electrode configuration [57]. The electrode was set up in such a way that the article incorporation into films occurs on either the interior rotating cylinder or the outer stationary cylinder [58, 59]. Afshar et al. [60] obtained composite coatings of bronze (90% Cu and 10% Sn) with 8.5% (vol.) graphite particles from an alkaline bath containing cyanide. A composite copper coating containing lubricating oil microcapsules was prepared by Zhu and Zhang [61], with emphasis on codeposition mechanisms and functions of the composite coating. Various types of particles can be incorporated into metallic or nonmetallic coatings. For example, SiC particles have been embedded into a Ni matrix with different electrical waveforms [62]. Particles of TiO_2 have been codeposited into lead for applications in lead-acid batteries [63]. Oxide and/or hydroxide particles have been codeposited into polymers or polyelectrolytes [64]. Submicron polymeric particles were electrocodeposited into copper as well [65].

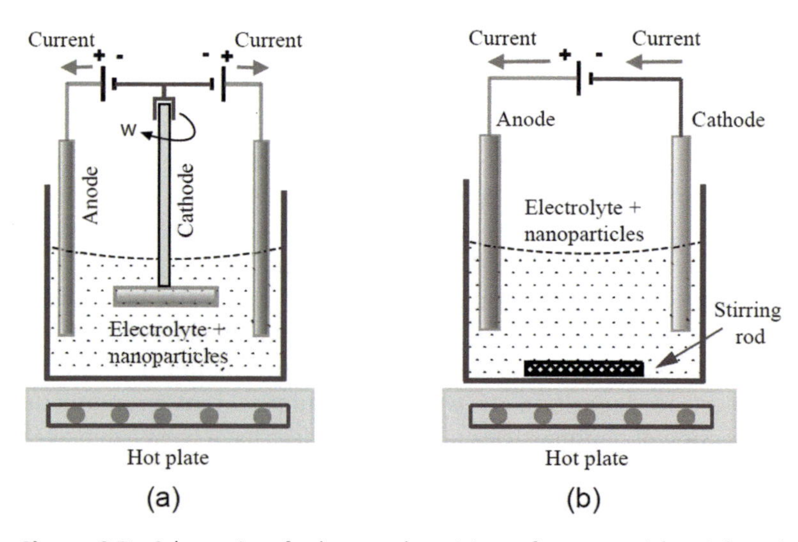

Figure 2.7 Schematic of electrocodeposition of nanoparticle-reinforced composites: (a) codeposition on a rotating cathode and (b) codeposition on a stationary cathode.

Electrocodeposition has been used for fabricating micro- and nanoscale devices with very high aspect ratios, such as micromembrane pulp systems [66] and microgears [67]. Another potential application of nanocomposite thin films is to protect against abrasive wear. Still another potential application is in mechanical resonator materials in radio frequency microelectromechanical systems (RF-MEMS). The main reliability issue in RF-MEMS is not fatigue or fracture but rather adhesion of a freestanding thin film to either an adjacent conductor or the substrate material [68]. This type of failure can be ameliorated by increasing the hardness and stiffness of the thin films [69].

The morphology and mechanical properties of electrocodeposited Al_2O_3 nanoparticle–reinforced copper matrix nanocomposites were investigated [70]. Before the electrode position, a 5 nm thick thin film of Cr was thermally evaporated on the substrate to promote good adhesion of the film to the substrate. A 50 nm thick gold coating was deposited on the Cr layer to provide an electrically conductive surface, with minimal oxidation, onto which the nanocomposite thin film can be subsequently deposited. To deposit a copper metal matrix nanocomposite thin film, an electrolyte of copper (II) sulfate ($CuSO_4{\cdot}5H_2O$) was made. The Al_2O_3 nanoparticles used had an average diameter of approximately 50 nm. The electrochemical cell for electrocodeposition of the copper and nanoparticles is shown in Fig. 2.7b. The anode is a pure copper plate, and the cathode is a silicon wafer with chromium and gold coatings. The grain size of the nanocomposite film is significantly smaller than that of the control films of pure copper. Electron backscatter diffraction (EBSD) experiments indicate that neither the nanocomposite thin films nor the control films exhibit a crystallographic texture [71]. Nanoindentation experiments show that the hardness of the nanocomposite thin film is approximately 25% higher than the hardness of the control films of pure copper. In addition, freestanding nanocomposite films removed from the silicon wafers were obtained and were used for static and fatigue tests [72].

2.5 Template Approaches

Template-based nanofabrication has been proven to be a versatile and simple approach for the preparation of uniform and ordered nanoscale structures. For example, electrochemically induced AAO membranes have very uniform pores whose diameters range from

10 to 200 nm and lengths from 1 to 100 μm depending on the preparation conditions. These cylindrical pores are well organized all over the membrane and run through the whole membrane. Such nanostructured AAO membranes have been used extensively to fabricate nanometer-scale fibrils, rods, belts, and wires of various materials [73–75]. In addition to AAO templates, other types of templates, such as PCs [76], polystyrene (PS) [77], and DNA [78], have also been used for nanofabrication.

2.5.1 AAO Template

There are many research papers available on AAO template preparation, for example, by Zhao et al. [79] and Kong [80]. The two-step anodic oxidization method was found to be better than the one-step approach because nanopores from two-step anodizing have a uniform size and a thin barrier layer can be obtained on the bottom of the pores. The typical procedure for the preparation of AAO templates via the two-step method is as follows. High-purity aluminum sheets are anodized on a single side using a regulated DC power supply in a 0.5 M oxalic acid solution. Before anodization, the aluminum plates need to be degreased in trichloroethylene for 2 h, followed by 10 min. of ultrasonic cleaning in acetone. Then the samples are rinsed first with methanol and then with distilled water. After that, the aluminum plates are etched in 5.0 M NaOH at 60°C for 20 min. and subsequently rinsed with distilled water. Electropolishing of the aluminum plates is conducted in a 30% HNO_3 methanol solution at –20°C. Anodizing is performed in 0.3 M $H_2C_2O_4$ at 0°C, held in an icy water bath. The cleaned and electropolished aluminum sheets are used as the anode and a Pt foil or a graphite plate as the cathode. The first anodization takes about 2 h. After the first anodization, the strip-off process should be carried out in a solution mixture of H_3PO_4 and H_2CrO_4. The exposed and well-ordered concave patterns on the aluminum substrate will act as a self-assembled mask for the second anodization process. The second anodization takes about 4 h. After the second anodization, AAO templates with uniform nanopores are obtained, as schematically shown in Fig. 2.8a.

The fabrication of a high-aspect-ratio nanofiber array using AAO templates was performed through a sol-gel approach [81].

The template used is a commercially available anodized AAO called Anodisc^{TM}, manufactured by Whatman International. The pore size of the AAO template is 200 nm. Deposition of NiO was conducted via the chemical reactions as shown in Eqs. 2.15a and 2.15b:

$$Ni^{2+} + 2OH^{-} = Ni(OH)_2 \quad (2.15a)$$

$$Ni(OH)_2 = NiO + H_2O \quad (2.15b)$$

The AAO template was put into a beaker containing 2M $NiSO_4{\cdot}H_2O$. The pH value of the solution was adjusted using an aqueous solution of sodium hydroxide. When the pH value reached about 12, $Ni(OH)_2$ gel formed within the AAO template. The template was then taken out from the beaker and dried to form nickel oxide fibers, as schematically shown in Fig. 2.8b.

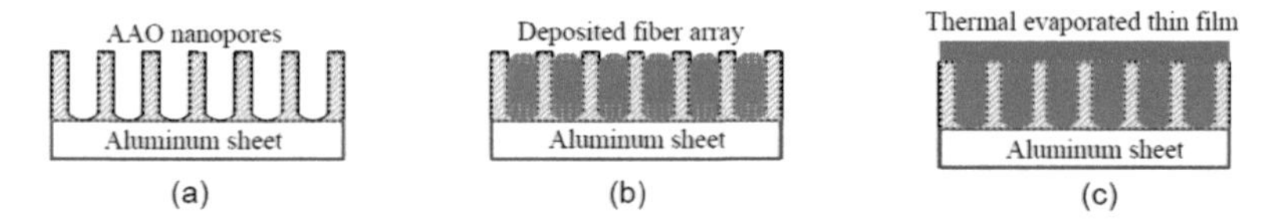

Figure 2.8 Preparation of AAO templates for nanofabrication: (a) preparation of AAO by anodizing aluminum, (b) deposition of the nanofiber array in the pores of the AAO template, and (c) assembling of the nanofiber array via thin-film technology.

Various types of 1D nanomaterials and structures have been synthesized. For example, Prussian blue NTs using AAO templates were prepared by Johansson et al. [82]. PS nanorods and NTs within cylindrical alumina nanopores were obtained by casting the polymer melts into the AAO templates [83]. Metallic nanofiber arrays in AAO templates may also be prepared through the electrodeposition technique as reviewed in Section 2.4.4. The nanofibers deposited in the AAO templates can be assembled by thermally evaporating thin films on the top surface of the templates to connect the fibers as illustrated in Fig. 2.8c. Once the AAO templates and aluminum sheets are dissolved by NaOH, well-aligned, freestanding nanofibers are obtained. Such nanoscale structures containing fiber arrays have many potential applications, as will be discussed in later sections.

2.5.2 Other Templates

In addition to AAO templates, there are many other types of templates used in nanofabrication. Zeolites have nanoporous structures that can accommodate guest materials in them. Thus, they have been considered as templates for nanofabrication. For example, $(CdP_2)_n$ nanoclusters were synthesized by incorporation into pores of zeolite Na-X and post-treated by laser ablation. The most stable clusters are $(CdP_2)_6$ and $(CdP_2)_8$ in the size region up to the dimension of the zeolite supercage [84]. Similar procedures were used to fabricate $(ZnAs_2)_n$ subnanoclusters [85] and ZnP_2 nanoparticles [86] in the Na-X zeolite. Unique optical properties of these $(CdP_2)_n$ nanoclusters, $(ZnAs_2)_n$ subnanoclusters, and ZnP_2 nanoparticles were found. Se_8-ring clusters were prepared by Lin et al. by loading Se into the cages of zeolite 5A [87]. In Poborchii's work [88], sulfur, selenium, and tellurium clusters confined in nanocavities of zeolite A were fabricated.

PC-based porous structures can be made by a track etching process [76]. It is shown that such nanostructures can be used as templates for nanofabrication of polymeric and metallic wires or tubules with interesting properties. For example, copper nanofiber arrays were fabricated in the pores of PC membranes [42]. Research on the development of templates from polyimide support was also reported in the literature [76]. It is noted that spherical polymers or ceramics can also be used as templates for the fabrication of nanomaterials. Yoon et al. [77] employed PS spheres as templates for the synthesis of silica mesoporous shells with highly ordered hexagonal crystalline structures, as confirmed by an XRD pattern and TEM image. The inner diameter of the shells is about 2.1 nm, and the surface area is as high as 1387 m^2/g.

2.6 Lithography

Lithography is a process in which a radiation-sensitive layer is applied to form predefined features on a surface. There are several methods associated with the lithography technique. Conventional photolithography, edge technique, soft lithography, electron-beam (E-beam) lithography, scanning probe technique, and X-ray lithography are some examples. In this part, a brief review of some of these methods will be provided. Nanofabrication via lithography originates from the conventional photolithographic technique. Thus, the basic procedures associated with photolithography are presented first, as follows.

2.6.1 Conventional Photolithography

In conventional photolithography, a photoresist (PR) layer is first spun on a substrate, for example, a Si wafer. The PR layer is baked and exposed to ultraviolet (UV) light through a mask. Typically, the mask is made of a quartz glass plate coated with a thin film of chromium. The chromium coating is selectively removed to form predefined patterns so that UV light can pass through some regions. The exposed PR is developed and leaves patterns on the substrate. If a positive PR is used, the exposed region is removed and the nonexposed region is kept upon development. On the contrary, if negative photoresist is used, the exposed region remains on the surface of the substrate after development. Figure 2.9 schematically shows the fabrication process for obtaining a microscale feature on a silicon wafer by photolithography.

Figure 2.9a shows the cleaned wafer to be processed. Figure 2.9b shows the SiO_2 coating on both sides of the Si wafer obtained through dry oxidization. The next step is to spin a positive PR on both sides of the wafer, as shown in Fig. 2.9c. In Fig. 2.9d, exposure of one side of the silicon wafer to the UV light passing through a mask is shown. After the silicon is exposed, the PR is selectively dissolved in a developer solution and the pattern is produced as shown in Fig. 2.9e. Figure 2.9f is the schematic of the wet-etched state of the silicon wafer. HF is used to remove the exposed SiO_2 following the reaction, as shown in Eq. 2.13d. The dissolving of SiO_2 in HF belongs to isotropic etching. Following the isotropic etching, the PR is stripped and the unetched SiO_2 region protected by the PR is exposed, as shown in Fig. 2.9g. The final step is wet etching in KOH to produce microscale patterns. The reaction mechanisms are shown in Eq. 2.16:

$$Si + 2OH^- = H_2SiO_3 + H_2\uparrow \tag{2.16}$$

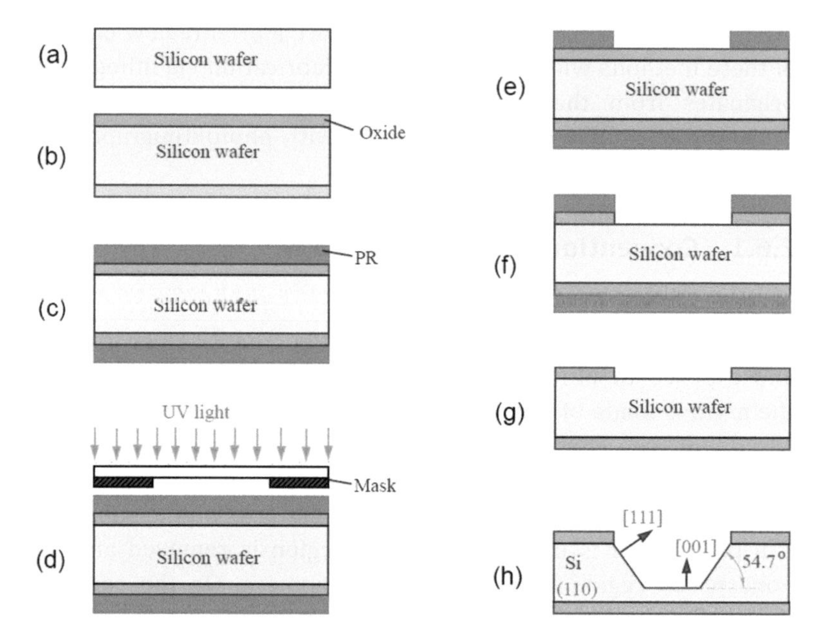

Figure 2.9 Schematic of conventional photolithography: (a) Si single-crystal wafer; (b) oxidization of the wafer to form a SiO_2 surface layer; (c) spin coating of the positive photoresistor; (d) UV exposure, (e) development, and (f) etching of the SiO_2 layer; (g) stripping off of the photoresistor; and (h) wet etching of the Si wafer.

The overall reaction as shown above consists of several steps [21, 22, 36]. The first step is the reaction of Si atoms with hydroxyl ions and release of electrons. The second step is the reduction of water and evolution of hydrogen gas. In the third step, the soluble silicon hydroxide (called sol) converts into a silicate gel and gradually precipitates from the sol. KOH etching is anisotropic. If the single-crystal Si (001) wafer is etched, the etching rate along the direction parallel to the <111> crystallographic direction is significantly lower than that along other directions. Thus, an etching pit with the defined orientation as shown in Fig. 2.9h is obtained.

2.6.2 Soft Lithography

Soft lithography employs an elastomer stamp or mold to form micro- and nanoscale features on a substrate. The commonly used elastomer for making the mode is polydimethylsiloxane (PDMS).

Fig. 2.10a shows such a PDMS mold with a defined pattern. In Fig. 2.10b, a prepolymer solution is poured into the cavity of the mold. Figure 2.10c shows a cleaned silicon wafer. The PDMS mold is impregnated with the prepolymer and pressed onto the surface of the wafer, while curing of the prepolymer occurs at elevated temperatures, as shown in Fig. 2.10d. After the mold is lifted off from the surface of the wafer, the defined polymer pattern is obtained as illustrated in Fig. 2.10e.

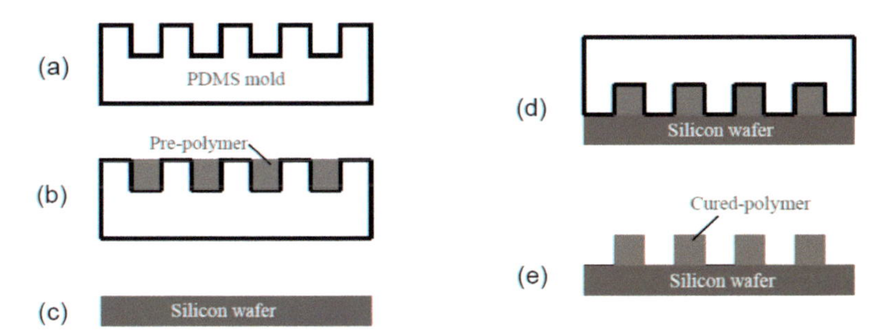

Figure 2.10 Soft lithography technique for nanofabrication: (a) PDMS mold, (b) impregnation of the prepolymer, (c) silicon wafer substrate, (d) molding and curing, and (e) pattern on the substrate.

Nanoimprinting is the application of soft lithography with other fabrication techniques. An example is given here to show the idea that well-aligned 1D nanoscale features can be fabricated through nanoimprinting. The first step is to immerse a specially designed PDMS mold with predefined patterns into a solution containing catalyst colloids, as shown in Fig. 2.11a. The catalyst is attached to the surface of the mold. Then the mold is stamped onto a silicon wafer, as shown in Fig. 2.11b. The catalyst is transferred onto the surface of the wafer and preserves the stamped pattern, as illustrated in Fig. 2.11c. In Fig. 2.11d, fabrication of nanofibers via catalyst-assisted growth is presented. More complicated nanoimprinting techniques have recently been developed. In the work performed by Kwon et al. [89], a mold-to-mold cross-imprinting (MTMCI) process, which redefines an imprint mold with another imprint mold, was studied. Fabrication of metallic nanodot arrays in a large area using the MTMCI process was carried out. By configuring two identical imprint molds with silicon spacer NWs in a perpendicular arrangement, a large array of silicon nanopillars with a dimension of less than

30 nm was obtained. Then, nanoimprint lithography with the silicon nanopillar mold was performed to fabricate arrays of Pt dots in large areas.

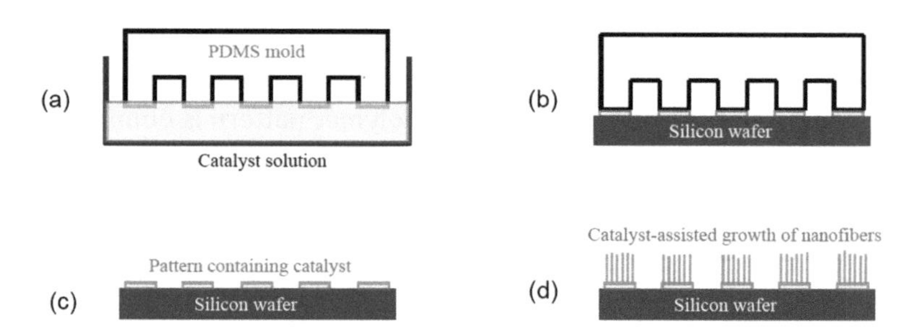

Figure 2.11 Nanoimprinting: (a) PDMS mold immersed in catalyst solution, (b) the mold stamped on a silicon wafer, (c) catalyst imprints on the wafer, and (d) catalyst-assisted growth of nanofiber arrays.

2.6.3 Near-Field Lithography

Near-field lithography is also called edge lithography or edge transfer lithography. The working mechanisms of near-field-edge lithographic patterning techniques are shown in Fig. 2.12. A special mask, called a phase mask, is used to generate the required edge effect. As shown in Fig. 2.12a, the phase mask is transparent to the UV light. The intensity of the light passing through both the thicker part and the thinner part is almost the same except for the locations at the edges. Owing to light interference, the intensity of the light at the edges is significantly lower than in the other parts, as shown in Fig. 2.12b. Figure 2.12c shows an oxidized Si wafer with spin-coated PR. If the PR is a positive one, the pattern developed is shown in Fig. 2.12d. On the contrary, a negative PR results in the pattern as shown in Fig. 2.12e.

From Fig. 2.12, it can be seen that edge lithographic patterning techniques have the capability of utilizing the edge effect from a mask with microscale patterns to generate nanoscale features. The phase mask may be a PDMS elastomeric stamp similar to that illustrated in Fig. 2.10a for soft lithography. In such a case, near-field lithography permits local surface modification in a single step through deposition of a SAM onto a metal substrate selectively along the feature edges of the PDMS elastomeric mold. For example, two types of stamps are designed to fabricate nanometer structures containing gold using edge transfer lithography [90]. It is also possible that modified stamps have the potential for producing very high resolution patterns with the versatility and simplicity of microcontact printing. Fabrication of biological nanostructures via the scanning near-field photolithography of a chloromethylphenylsiloxane (CMPS) monolayer has been investigated [91]. DNA surface patterns with dimensions less than 100 nm were obtained using a near-field scanning optical microscope coupled to a UV laser and a CMPS SAM. The process involves two steps. First, the CMPS SAM is exposed to the 244 nm UV light to create nanoscale patterns of surface carboxylic acid functional groups. Then, the carboxylic acid functional groups convert to *N*-hydroxysuccinimidyl ester and react with DNA to spatially control DNA grafting with high selectivity.

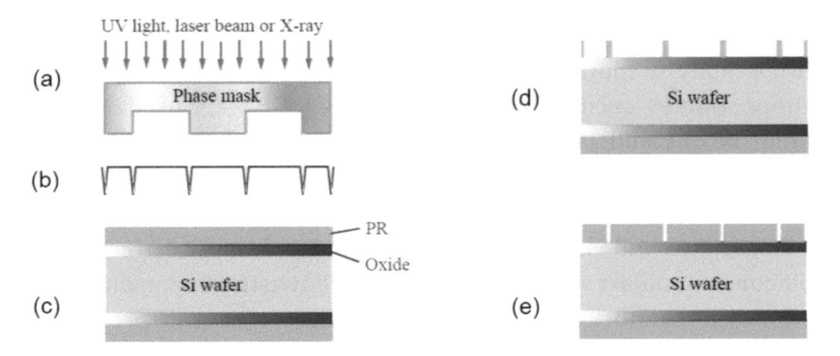

Figure 2.12 Edge techniques in photolithography: (a) UV radiation passing through the phase mask, (b) intensity of the transmission light, (c) UV exposure of photoresistors on the oxide layer of a Si wafer, (d) pattern formation using a positive photoresistor, and (e) a pattern obtained using a negative photoresistor.

2.6.4 Electron-Beam Lithography

E-beam lithography uses a focused electron beam as the irradiation for exposure of PRs such as polymethyl methacrylate (PMMA). PMMA is a positive resist that can a provide high contrast and a high resolution for E-beam lithographic processes. A beam blocker or shutter, as shown in Fig. 2.13a, may also be used to reduce electron scattering and assist in forming patterns. Figure 2.13b shows the developed pattern on the silicon wafer. The pattern can be as fine as 10 nm. Applying deep reactive ion etching (DRIE) of the wafer results in vertical nanopores within the silicon wafer, as shown in Fig. 2.13c. The developed pattern in Fig. 2.13b may be covered by a thermally evaporated Au thin film, as shown in Fig. 2.13d. After the PMMA is stripped off, a Au pattern on the silicon wafer is obtained (see Fig. 2.13e). Such a nanoscale Au pattern may be used as the catalyst for the fabrication of 1D nanomaterials. Figure 2.13f illustrates the laser- and catalyst-assisted growth of nanofibers with heterogeneous structures as mentioned in Section 4.2. During E-beam lithography, the beam bombards PRs point by point. Thus, it is difficult to obtain comparable high rates in assembly-line manufacturing processes. There are some other high-energy-beam-related fabrication processes that are similar to E-beam lithography. X-ray lithography is one of these. The resolution of X-ray lithography is 20 nm. Owing to the complexity of mask technology and exposure systems, X-ray lithography is more expensive for practical applications [20, 92]. Other sources, such as focused ion beams [93], were used. Neutral atom beams and laser beams may also be used for exposure in nanofabrication [20].

Juhasz and Linnros [94] performed nanofabrication by E-beam lithography and laser-assisted electrochemical size reduction. Silicon nanopillars were reactively etched down to a 10 nm diameter, while the cylindrical shape of these pillars was well preserved. Selective etching of different parts of the pillars by varying the applied bias voltage was also demonstrated. Variable pressure E-beam lithography was used to directly pattern nanoscale features on substrates with low electrical conductivity [95]. This approach has the high-resolution patterning capability of E-beam lithography. It also has the charge-balance mechanism of the variable pressure scanning electron microscope to control charging effects during pattern exposure. Thus, it is not necessary to use any additional materials or processing steps to eliminate pattern distortion during high-vacuum E-beam patterning on such substrates. The shape of the scattering profile in the resist layer is modified in the presence of the chamber gas, allowing improved pattern definition at higher pressures.

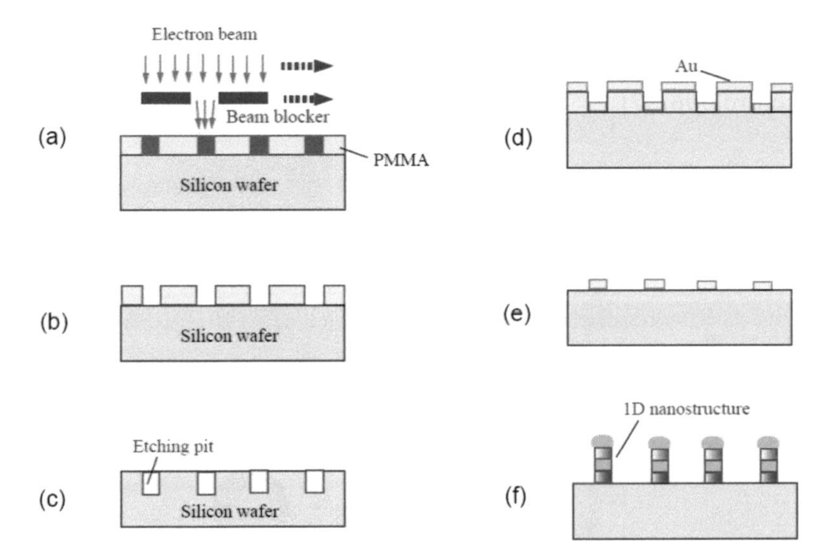

Figure 2.13 Electron-beam lithography technique for nanofabrication: (a) PMMA exposure under a scanning electron beam, (b) pattern development, (c) generation of nanoscale pores by DRIE, (d) an evaporating Au thin film on PMMA, (e) stripping off of PMMA to form an Au pattern on a Si substrate, and (f) laser- and catalyst-assisted growth of 1D nanostructures.

2.6.5 Scanning Probe Lithography

Scanning probe microscopy (SPM) can be used for nanoscale manipulation such as nanoscale assembling and nanolithography [20]. Figure 2.14 demonstrates the idea of using the SPM for nanoscale assembling. Metallic nanoparticles such as gold and silver particles can be moved under the drag of the SPM tip. The formed patterns can be used as the masks for nanolithography. In addition to metals, substances containing molecules with multifunctional groups such as hydrophilic and hydrophobic groups in the form of menisci can be assembled by SPM. Such a scanning probe lithographic technique is called dip-pen lithography. Comprehensive experimental and molecular dynamics studies have been performed to understand the effect of molecular type, concentration, temperature, voltage, and the gap between the probe tip and the substrate (such as a silicon wafer) on the quality of the self-organized nanoscale patterns. Explicit relationships are established between the properties of material and the processing variables. Reverse analysis is also employed to obtain desired patterns by material selection and fabrication parameter control [96, 97].

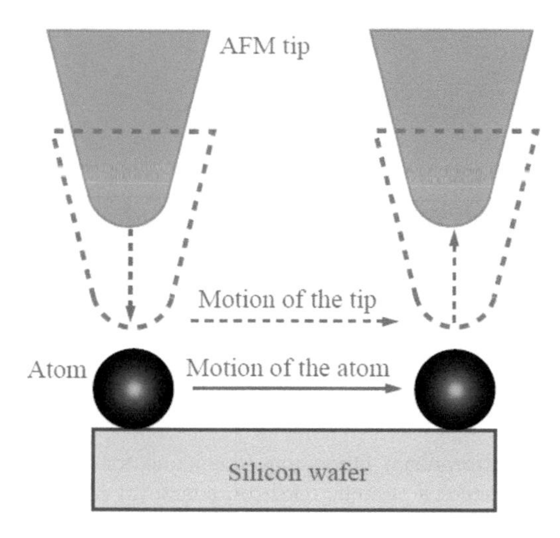

Figure 2.14 Scanning probe technique of nanofabrication.

SPM has been used as a nanometer-scale lithographic tool for electrode surface modification [98]. An organosilane monolayer composed of octadecylsilyl [$CH_3(CH_2)_{17}Si$] groups was prepared by CVD and served as a resist material for scanning probe lithography. Using an atomic force microscope (AFM) with an electrically conductive probe, the monolayer deposited on a Si substrate was patterned by allowing current to flow through the AFM-probe monolayer junction. It is noted that the density of the tunneling current from the AFM tip is determined by several factors, such as the property of the monolayer junction and the gap between the AFM tip and the monolayer [20]. The pattern was formed on the monolayer and was transferred to the substrate of Si by chemical etching in an aqueous solution mixture of NH_4F and H_2O_2. The etching proceeded in the regions where the probe had passed. The etching process is area selective because the monolayer had been degraded electrochemically in these regions. In the same paper, Sugimura et al. [98] also introduced a nanofabrication method that uses the patterned organosilane monolayer as a template for pattern transfer to an Au nanostructure. The patterned organosilane monolayer was first etched in a HF solution in order to expose the substrate Si in the monolayer-degraded region. The HF-etched sample was next treated in an Au electrode plating bath. In this plating, deposition of Au proceeded selectively on the exposed Si area, while the surrounding undegraded monolayer surface remained free of deposits. The probe-scanned pattern was transferred to an Au pattern through this area-selective electrode plating.

Selective chemical processing down to the atomic scale on hydrogen-passivated silicon surfaces was demonstrated using ultrahigh-vacuum scanning tunneling microscopy [99]. The difference in chemical properties between clean and H-passivated Si(100) makes selective chemical processing such as nanoscale-selective oxidation, nitridation, molecular functionalization, and metallization possible. Current sensing AFM was used to form nanopatterns of alkanethiol SAMs on a gold substrate [100]. It was found that when the conducting tip with a sufficient bias was used to scan over the SAM in a toluene solution containing alkanethiol of a different chain length, the original SAM underneath the tip was removed and a new nanospace-confined SAM of the alkanethiol in toluene was formed. The existence of the nanopatterns was confirmed by the variation in conductivity and current-bias characteristics.

Dip-pen nanolithography (DPN) is the technique for creating nanoscale chemical patterns on substrates by using the probe of a scanning probe microscope or an AFM. In DPN, molecules in liquid droplets can be precisely placed. It is suitable for use in nanomanipulation. However, the successful use of this technique is strongly dependent on different variables, including the wetting conditions of drops on the surface of scanning probe microscopic tips. A model was developed to predict the shapes and locations of axisymmetric; equilibrium drops on conical, solid surfaces as a function of drop volume; needle geometry and shape; needle surface wettability (contact angle); liquid surface tension; line tension; and gravity [101]. Calculation of time-dependent spreading of liquid on substrates was also performed [102]. The fabrication of luminescent nanostructures via DPN was shown by Noy et al. [103]. A combination of DPN and scanning optical confocal microscopy was used to fabricate and visualize various materials, including proteins, in the form of nanoscale patterns on glass substrates. Polymer NWs of controlled size using conductive polymers as start materials were also obtained. Some important factors that influence the size of these NWs were identified as well. Through DPN, it is also possible to position protein molecules [104] or Fe_3O_4 magnetic particles [105] on DNA chains.

Improvement on DPN technique is underway. To increase the resolution of nanopatterns, the effects of changing the tip radius and surface roughness were investigated [106]. The results helped to improve the performance of a facility called the Nscriptor DPN instrument. The instrument with an enhanced resolution demonstrates the placement of pattern features with precision better than 10 nm. The idea of resistless photolithography using laser for the fabrication of microscopic markers and electrodes for DPN was proposed [107]. The electrodes and markers are robust and can withstand harsh chemical treatments. Demonstration on gold colloid nanopatterning using electrodes and markers was provided. Electrostatic actuation to create an active DPN probe array was performed by Bullen and Liu [108]. It is reported that electrostatic actuation has the advantage of actuation without the probe being heated, as in the case of thermal bimetallic actuation. Another advantage is that more densely spaced probe arrays can be built because the actuator cross talk between the neighboring probes is reduced.

2.7 Electrospinning

Electrospinning is the process during which forced viscous flowing of polymer solutions from a jet to a target occurs. The driving force for the flow is mainly electrical. Pressure may also be applied to the solution to facilitate the viscous flow. The stretch of molecular chains occurs simultaneously. The jet and the target are attached to specially designed electrodes with applied high voltages. A simple electrospinning system is shown in Fig. 2.15a. It consists of a unit for supplying the charged polymer solution (or melt), a collecting plate, and a 10–15 kV DC power source. The collecting plate may be a metallic screen, a rotating disc, or a mandrel. During electrospinning, the solvent gradually evaporates and a charged polymer fiber is left to accumulate on the target. It must be pointed out that nanoparticles may be added to polymer solutions so that the spun products are composite fibers. For example, polyacrylonitrile (PAN)-Fe_3O_4 nanocomposite fibers were made by Luoh and Hahn [109]. The charge on the fiber dissipates into the surrounding environment. In some cases, a UV or infrared (IR) light source may be attached to the system to assist in the curing of the polymer fibers. Typical resulting products include fiber filaments and nonwoven fiber mats consisting of micro- or nanoscale fibers. The size, orientation, and properties of the electrospun fibers can be varied by changing the associated processing parameters [110]. Electrospinning has provided a simple approach to fabricating nanofibers and assemblies with controllable structures. In Fig. 2.15b, a SEM image of the woven PAN nanofibers made in our lab is shown.

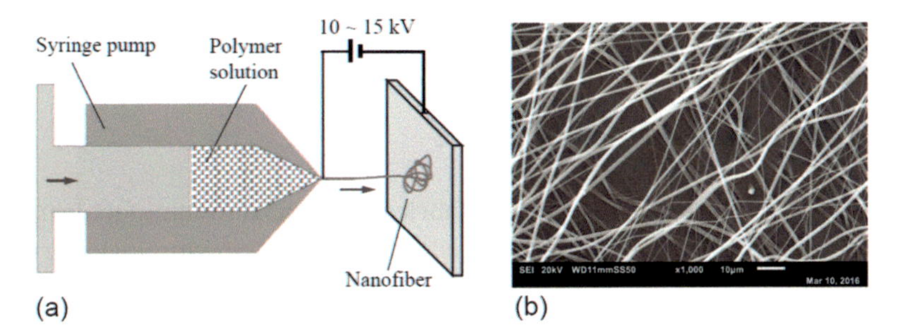

Figure 2.15 Electrospinning process and the morphology of the manufactured fibers: (a) schematic of the setup and (b) morphology of the electrospun fibers.

In addition to polymeric fibers, ceramic fibers can be obtained by electrospinning combined with calcination [111, 112]. For example, the spinel $NiCo_2O_4$ nanofibers with diameters of 50–100 nm were prepared [111]. The fabrication process includes electrospinning of the PVA–cobalt acetate–nickel acetate composite precursor to form an inorganic-polymer composite precursor, followed by high-temperature calcination of the precursor to produce the $NiCo_2O_4$ nanofibers. V_2O_5 nanorods on TiO_2 nanofibers were also prepared by electrospinning and calcination [112]. Single-crystal V_2O_5 nanorods were grown on rutile nanofibers by carefully calcining composite nanofibers consisting of amorphous V_2O_5, amorphous TiO_2, and polyvinylpyrrolidone (PVP). The dimension of the synthesized V_2O_5

nanorods is dependent on the composition of the nanofibers and the processing parameters, such as the calcination temperature. It is expected that this approach can also be extended to fabricate other, more complex micro- or nanoscale architectures. Near-field electrospinning has been developed to fabricate solid nanofibers in a controllable and direct way [113]. A tungsten electrode with a sharp tip of 50 μm diameter was used as a dip pen to deliver a liquid polymer solution so that nanofibers 50–500 nm in size were obtained on silicon-based collectors. The tungsten electrode-to-collector distance can be set to a gap as small as 500 μm to achieve position-controllable deposition. This is much less than the distance used in ordinary electrospinning facilities, which is above 15 mm. In addition, the applied bias voltage is much lower than that used in conventional electrospinning. Therefore, near-field electrospinning is promising as a technique for directly fabricating various nanofibers.

2.8 Concluding Remarks

In this chapter a number of nanofabrication and nanomanufacturing techniques have been presented, showing that the technologies are under fast development. It is noted that significant progress has been made in the field. The need for new nanomaterials, devices, and systems with novel functions becomes the momentum for improving the existing nanofabrication techniques and exploring new ones, which open numerous opportunities of research, development, and education in the field. For example, the shortage of energy pushes us to find clean and sustainable energy sources and to design and manufacture systems with high energy efficiency. Nanofabrication and nanomanufacturing are expected to play an important role in resolving this issue. Fabricating low-cost and high-efficiency TE nanosystems to convert waste heat into storable electricity is considered to be one of the solutions for energy saving. Another example to show the critical role of nanofabrication is photocatalytic reduction of carbon dioxide into methanol. Novel nanoreactors may make this possible, and this technology would be doubly beneficial to our society in the sense of ameliorating environmental pollution and helping create a kind of clean fuel. To make such nanoreactors practical for carbon dioxide reduction, there still remain a lot of challenges in nanofabrication and nanomanufacturing technologies. To meet such challenges, one has to build a solid foundation in the field first. Owing to the interdisciplinary nature of nanofabrication and nanomanufacturing, to understand the fundamentals an individual needs to have knowledge of physics, chemistry, materials science, and so on. To obtain significant achievement in the field, the cooperation of scientists, engineers, and educators from different fields is indispensable.

References

1. Sarafraz, M. M., Hormozi, F. (2016). Heat transfer, pressure drop and fouling studies of multi-walled carbon nanotube nano-fluids inside a plate heat exchanger, *Exp. Therm. Fluid Sci.*, **72**(4), pp. 1–11.
2. Sarafraz, M. M., Hormozi, F., Silakhori, M., Peyghambarzadeh, S. M. (2016). On the fouling formation of functionalized and non-functionalized carbon nanotube nano-fluids under pool boiling condition, *Appl. Therm. Eng.*, **95**(2), pp. 433–444.
3. Kousalya, A. S., Singh, K. P., Fisher, T. S. (2015). Heterogeneous wetting surfaces with graphitic petal-decorated carbon nanotubes for enhanced flow boiling, *Int. J. Heat Mass Transfer*, **87**(8), pp. 380–389.
4. Zhang, T., Wu, S., Xu, J., Zheng, R., Cheng, G. (2015). High thermoelectric figure-of-merits from large-area porous silicon nanowire arrays, *Nano Energy*, **13**(4), pp. 433–441.
5. Lozano, K., Hernandez, C., Petty, T. W., Sigman, M. B., Korgel, B. (2006). Electrorheological analysis of nano laden suspensions, *J. Colloid Interface Sci.*, **297**(2), pp. 618–624.
6. Sysoev, V. V., Button, B. K., Wepsiec, K., Dmitriev, S., Kolmakov, A. (2006). Toward the nanoscopic “electronic nose”: hydrogen vs carbon monoxide discrimination with an array of individual metal oxide nano- and mesowire sensors, *Nano Lett.*, **6**(8), pp. 1584–1588.
7. Tang, X., Bansaruntip, S., Nakayama, N., Yenilmez, E., Chang, Y., Wang, Q. (2006). Carbon nanotube DNA sensor and sensing mechanism, *Nano Lett.*, **6**(8), pp. 1632–1636.
8. Robinson, J. A., Snow, E. S., Bdescu, S. C., Reinecke, T. L., Perkins, F. K. (2006). Role of defects in single-walled carbon nanotube chemical sensors, *Nano Lett.*, **6**(8), pp. 1747–1751.

9. Hwang, S., Jeong, S., Lee, O., Lee, K. (2005). Fabrication of vacuum tube arrays with a sub-micron dimension using anodic aluminum oxide nano-templates, *Microelectron. Eng.*, **77**(1), pp. 2–7.
10. Sohn, J. I., Lee, S., Song, Y., Choi, S., Cho, K., Nam, K. (2006). Large field emission current density from well-aligned carbon nanotube field emitter arrays, *Curr. Appl. Phys.*, **1**(1), pp. 61–65.
11. Chueh, Y., Ko, M., Chou, L., Chen, L., Wu, C., Chen, C. (2006). $TaSi_2$ nanowires: a potential field emitter and interconnect, *Nano Lett.*, **6**(8), pp. 1637–1644.
12. Zhang, X., Liu, H., He, W., Wang, J., Li, X., Boughton, R. I. (2004). Synthesis of monodisperse and spherical YAG nanopowder by a mixed solvothermal method, *J. Alloys Compd.*, **372**(1–2), pp. 300–303.
13. Yamanaka, S., Hamaguchi, T., Muta, H., Kurosaki, K., Uno, M. (2004). Fabrication of oxide nanohole arrays by a liquid phase deposition method, *J. Alloys Compd.* **373**(1–2), pp. 312–315.
14. Cavallini, M., Facchini, M., Massi, M., Biscarini, F. (2004). Bottom–up nanofabrication of materials for organic electronics, *Synth. Met.*, **146**(3), pp. 283–286.
15. Cao, H. L., Qian, X. F., Gong, Q., Du, W. M., Ma, X. D., Zhu, Z. K. (2006). Shape- and size-controlled synthesis of nanometre ZnO from a simple solution route at room temperature, *Nanotechnology*, **17**(15), pp. 3632–3636.
16. Carny, O., Shalev, D. E., Gazit, E. (2006). Fabrication of coaxial metal nanocables using a self-assembled peptide nanotube scaffold, *Nano Lett.*, **6**(8), pp. 1594–1597.
17. Hou, S., Harrell, C. C., Trofin, L., Kohli, P., Martin, C. R. (2004). Layer-by-layer nanotube template synthesis, *J. Am. Chem. Soc.*, **126**(18), pp. 5674–5675.
18. Shin, H., Jeong, D. K., Lee, J., Sung, M. M., Kim, J. (2004). Formation of TiO_2 and ZrO_2 nanotubes using atomic layer deposition with ultraprecise control of the wall thickness, *Adv. Mater.*, **16**(14), pp. 1197–1200.
19. Hsu, M. C., Leu, I. C., Sun, Y. M., Hon, M. H. (2005). Fabrication of CdS@ TiO_2 coaxial composite nanocables arrays by liquid-phase deposition, *J. Cryst. Growth*, **285**(4), pp. 642–648.
20. Poole, C. P., Owens, F. J. (2003). *Introduction to Nanotechnology*, New York: John Wiley.
21. Allen, J. J. (2005). *Micro Electro Mechanical System Design*, Boca Raton, FL: CRC Press.
22. Senturia, S. D. (2001). *Microsystem Design*, New York: Springer.

23. Goerigk, G., Williamson, D. L. (1998). Nanostructured Ge distribution in α-SiGe: H alloys from anomalous small-angle X-ray scattering studies, *Solid State Commun.*, **108**(7), pp. 419–424.
24. Meyyappan, M. (2005). *Carbon Nanotubes Science and Applications*, Boca Raton, FL: CRC Press.
25. Saito, R., Dresselhaus, G., Dresselhaus, M. S. (1998). *Physical Properties of Carbon Nanotubes*, London: Imperial College Press.
26. Xu, S., Ostrikov, K., Long, J. D., Huang, S. Y. (2006). Integrated plasma-aided nanofabrication facility: operation, parameters, and assembly of quantum structures and functional nanomaterials, *Vacuum*, **80**(6), pp. 621–630.
27. Wang, W. H., Chao, K. M., Teng, I. J., Kuo, C. T. (2006). Nanofabrication and the structure-property analyses of phase-change alloy-ended CNTs, *Surf. Coat. Technol.*, **200**(10), pp. 3206–3210.
28. Wang, W. H., Lin, Y. T., Kuo, C. T. (2005). Nanofabrication and properties of the highly oriented carbon nanocones, *Diamond Relat. Mater.*, **14**(3–7), pp. 907–912.
29. Wang, Y., Lew, K., Ho, T., Pan, L., Novak, S. W., Dickey, E. C., Redwing, J. M., Mayer, T. S. (2005). Use of phosphine as an n-type dopant source for vapor-liquid-solid growth of silicon nanowires, *Nano Lett.*, **5**(11), pp. 2139–2143.
30. Adhikari, H., Marshall, A. F., Chidsey, C. E. D., McIntyre, P. C. (2006). Germanium nanowire epitaxy: shape and orientation control, *Nano Lett.*, **6**(2), pp. 318–323.
31. Sköld, N., Karlsson, L. S., Larsson, M. W., Pistol, M., Seifert, W., Trägârdh, J., Samuelson, L. (2005). Growth and optical properties of strained GaAs-$Ga_xIn_{1-x}P$ core-shell nanowires, *Nano Lett.*, **5**(10), pp. 1943–1947.
32. Vaddiraju, S., Mohite, A., Chin, A., Meyyappan, M., Sumanasekera, G., Alphenaar, B. W., Sunkara, M. K. (2005). Mechanisms of 1D crystal growth in reactive vapor transport: indium nitride nanowires, *Nano Lett.*, **5**(8), pp. 1625–1631.
33. Wu, Y., Fan, R., Yang, P. (2002). Block-by-block growth of single-crystalline Si/SiGe superlattice nanowires, *Nano Lett.*, **2**(2), pp. 83–86.
34. Murray, B. J., Li, Q., Newberg, J. T., Menke, E. J., Hemminger, J. C., Penner, R. M. (2005). Shape- and size-selective electrochemical synthesis of dispersed silver(I) oxide colloids, *Nano Lett.*, **5**(11), pp. 2319–2324.
35. Mardilovich, P., Kornilovitch, P. (2005). Electrochemical fabrication of nanodimensional multilayer films, *Nano Lett.*, **5**(10), pp. 1899–1904.

36. Franssila, S. (2004). *Introduction to Microfabrication*, New York: John Wiley.

37. Gargas, D. J., Bussian, D. A., Buratto, S. K. (2005). Investigation of the connectivity of hydrophilic domains in nafion using electrochemical pore-directed nanolithography, *Nano Lett.*, **5**(11), pp. 2184–2187.

38. Sato, H., Homma, T., Mori, K., Osaka, T., Shoji, S. (2005). Picoliter volume glass tube array fabricated by Si electrochemical etching process, *Electrochim. Acta*, **51**(5), pp. 844–848.

39. Gros-Jean, M., Solomon, I., Chazalviel, J. N. (1999). Anodization of undoped amorphous silicon by electrical injection of holes, *Solid State Commun.*, **109**(10), pp. 643–648.

40. Dimova-Malinovska, D., Janvier, C., Sendova-Vassileva, M., Kamenova, M., Marinova, T., Krastev, V. (1996). Correlation between the photoluminescence and chemical bonding in porous silicon, *Solid State Commun.*, **99**(9), pp. 641–644.

41. Pesika, N. S., Radisic, A., Stebe, K. J., Searson, P. C. (2006). Fabrication of complex architectures using electrodeposition into patterned self-assembled monolayers, *Nano Lett.*, **6**(5), pp. 1023–1026.

42. Kaur, R., Verma, N. K., Thapar, S. K., Chakarvarti, S. K. (2006). Fabrication of copper microcylinders in polycarbonate membranes and their characterization, *J. Mater. Sci.*, **41**(12), pp. 3723–3728.

43. Stojak, J. L., Fransaer, J., Talbot, J. B. (2002). In *Advances in Electrochemical Science and Engineering* (R. C. Alkire and D. M. Kolb, eds.), Vol. 7, p. 193, Weinheim, Germany: Wiley-VCH.

44. Kuo, S. L., Chen, Y. C., Ger, M. D., Hwu, W. H. (2004). Nano-particles dispersion effect on Ni/Al_2O_3 composite coatings, *Mater. Chem. Phys.*, **86**(1), pp. 5–10.

45. Ahmad, Y. H., Mohamed, A. M. A. (2014). Electrodeposition of nanostructured nickel-ceramic composite coatings: a review, *Int. J. Electrochem. Sci.*, **9**(4), pp. 1942–1963.

46. Ozkan, S., Hapci, G., Orhan, G., Kazmanli, K. (2013). Electrodeposited Ni/SiC nanocomposite coatings and evaluation of wear and corrosion properties, *Surf. Coat. Technol.*, **232**(10), pp. 734–741.

47. Tripathi, M. K., Singh, D. K., Singh, V. B. (2013). Electrodeposition of Ni-Fe/BN nano-composite coatings from a non-aqueous bath and their characterization, *Int. J. Electrochem. Sci.*, **8**(3), pp. 3454–3471.

48. Mohan, S., Saravanan, G., Bund, A. (2012). Role of magnetic forces in pulse electrochemical deposition of Ni-nano Al(2)O(3) composites, *Electrochim. Acta*, **64**(3), pp. 94–99.

49. Qu, N. S., Chan, K. C., Zhu, D. (2004). Pulse co-electrodeposition of nano Al_2O_3 whiskers nickel composite coating, *Scr. Mater.*, **50**(8), pp. 1131–1134.

50. Boccaccini, A. R., Zhitomirsky, I. (2002). Application of electrophoretic and electrolytic deposition techniques in ceramics processing, *Curr. Opin. Solid State Mater. Sci.*, **6**(3), pp. 251–260.

51. Thiemig, D., Bund, A., Talbot, J. B. (2009). Influence of hydrodynamics and pulse plating parameters on the electrocodeposition of nickel-alumina nanocomposite films, *Electrochim. Acta*, **54**(9), pp. 2491–2498.

52. Sangeetha, S., Kalaignan, G. P. (2015). Tribological and electrochemical corrosion behavior of Ni-W/BN (hexagonal) nano-composite coatings, *Ceram. Int. (Part A)*, **41**(9), pp. 10415–10424.

53. Kartal, M., Uysal, M., Gul, H., Alp, A., Akbulut, H. (2015). Pulse electrocodeposition of Ni/MWCNT nanocomposite coatings, *Surf. Eng.*, **31**(9), pp. 659–665.

54. Kartal, M., Gul, H., Uysal, M., Alp, A., Akbulut, H. (2015). Ni/MWCNT coatings produced by pulse electrocodeposition technique, *Fullerenes Nanotubes Carbon Nanostruct.*, **23**(11), pp. 956–960.

55. Mohajeri, S., Dolati, A., Ghorbani, M. (2015). The influence of pulse plating parameters on the electrocodeposition of Ni-TiO_2 nanocomposite single layer and multilayer structures on copper substrates, *Surf. Coat. Technol.*, **262**(1), pp. 173–183.

56. Groza, J. R., Gibeling, J. C. (1993). Principles of particle selection for dispersion-strengthened copper, *Mater. Sci. Eng. A*, **171**(1–2), pp. 115–125.

57. Stojak, J. L., Talbot, J. B. (2001). Effect of particles on polarization during electrocodeposition using a rotating cylinder electrode, *J. Appl. Electrochem.*, **31**(5), pp. 559–564.

58. Talbot, J. B. (2004). Electrocodeposition of nanocomposite films, *Plat. Surf. Finish.*, **91**(10), pp. 60–65.

59. Stojak, J. L., Talbot, J. B. (1999). Investigation of electrocodeposition using a rotating cylinder electrode, *J. Electrochem. Soc.*, **146**(12), pp. 4504–4513.

60. Afshar, A., Ghorbani, M., Mazaheri, M. (2004). Electrodeposition of graphite-bronze composite coatings and study of electroplating characteristics, *Surf. Coat. Technol.*, **187**(2–3), pp. 293–299.

61. Zhu, L. Q., Zhang, W. (2004). The codeposition mechanism and function of oil-containing microcapsule composite copper coating, *Acta Phys. Chim. Sin.*, **20**(8), pp. 795–800.

62. Hu, F., Chan, K. C. (2004). Electrocodeposition behavior of Ni–SiC composite under different shaped waveforms, *Appl. Surf. Sci.*, **233**(1–4), pp. 163–171.

63. Barkleit, G., Grahl, A., Maccagni, M., Olper, M., Scharf, P., Wagner, R. (1999). Electrodeposited, dispersion-hardened, lightweight grids for lead–acid batteries, *J. Power Sources*, **78**(1–2), pp. 73–78.

64. Bon, P., Zhitomirsky, I., Embury, J. D. (2004). Electrodeposition of composite iron oxide–polyelectrolyte films, *Mater. Chem. Phys.*, **86**(1), pp. 44–50.

65. Eng, Y. (1991). Electrocodeposition of metal and colloidal particle composite films onto a rotating cylinder electrode, PhD thesis, Columbia University, New York.

66. Muller, R., Schmid, P., Munding, A., Gronmaier, R., Kohn, E. (2003). Elements for surface microfluidics in diamond, *Diamond Relat. Mater.*, **13**(4–8), pp. 780–784.

67. Yang, N. Y. C., Headley, T. J., Kelly, J. J., Hruby, J. M. (2004). Metallurgy of high strength Ni–Mn microsystems fabricated by electrodeposition, *Scr. Mater.*, **51**(8), pp. 761–766.

68. Rebeiz, G. M. (2003). *RF MEMS: Theory, Design and Technology*, Hoboken, NJ: Wiley.

69. Zhao, Y. P., Wang, L. S. (2003). Mechanics of adhesion in MEMS: a review, *J. Adhes. Sci. Technol.*, **17**(4), pp. 519–546.

70. Gan, Y., Lee, D., Chen, X., Kysar, J. W. (2005). Structure and properties of electrocodeposited Cu-Al_2O_3 nanocomposite thin films, *J. Eng. Mater. Technol.*, **127**(4), pp. 451–456.

71. Gan, Y. X. (2005). High strain gradient plasticity associated with the deformation of face-centered cubic single crystals, PhD thesis, Columbia University, New York.

72. Gan, Y. X., Wei, C. S., Lam, M., Wei, X. D., Lee, D. Y., Kysar, J. W., Chen, X. (2006). Deformation and fracture behavior of electrocodeposited alumina nanoparticle/copper composite films, *J. Mater. Sci.*, **42**(13), pp. 5256–5263.

73. Wan, D., Wang, Y., Zhou, Z., Yang, G., Wang, B., Wei, L. (2005). Fabrication of the ordered FeS_2 (pyrite) nanowire arrays in anodic aluminum oxide, *Mater. Sci. Eng. B*, **122**(2), pp. 156–159.

74. Xu, H. B., Chen, H. Z., Xu, W. J., Wang, M. (2005). Fabrication of organic copper phthalocyanine nanowire arrays via a simple AAO template-based electrophoretic deposition, *Chem. Phys. Lett.*, **412**(4–6), pp. 294–298.

75. Liu, L., Zhao, C., Zhao, Y., Jia, N., Zhou, Q., Yan, M., Jiang, Z. (2005). Characteristics of polypyrrole (PPy) nano-tubules made by templated ac electropolymerization, *Eur. Polym. J.*, **41**(9), pp. 2117–2121.

76. Ferain, E., Legras, R. (2003). Track-etch templates designed for micro- and nanofabrication, *Nucl. Instrum. Methods Phys. Res. B*, **208**(8), pp. 115–122.

77. Yoon, S. B., Kim, J. Y., Kim, J. H., Park, S. G., Kim, J. Y., Lee, C. W., Yu, J. S. (2006). Template synthesis of nanostructured silica with hollow core and mesoporous shell structures, *Curr. Appl. Phys.*, **6**(6), pp. 1059–1063.

78. Wood, J. (2006). DNA tiles set pattern for assembly: bionanotechnology, *Nano Today*, **1**(1), p. 9.

79. Zhao, Y., Chen, M., Zhang, Y., Xu, T., Liu, W. (2005). A facile approach to formation of through-hole porous anodic aluminum oxide film, *Mater. Lett.*, **59**(1), pp. 40–43.

80. Kong, L. B. (2005). Synthesis of Y-junction carbon nanotubes within porous anodic aluminum oxide template, *Solid State Commun.*, **133**(8), pp. 527–529.

81. Hyers, M. L., Gan, Y. X., Wei, C. S., Lewis, P. A., Flynn, G., Zhong, W. H. (2007). Morphology and deformation state of nanofibers in anodic aluminum oxide (AAO) templates, *J. Comput. Theor. Nanosci.*, **4**(1), pp. 111–121.

82. Johansson, A., Widenkvist, E., Lu, J., Boman, M., Jansson, U. (2005). Fabrication of high-aspect-ratio Prussian blue nanotubes using a porous alumina template, *Nano Lett.*, **5**(8), pp. 1603–1606.

83. Zhang, M., Dobriyal, P., Chen, J., Russell, T., P., Olmo, J., Merry, A. (2006). Wetting transition in cylindrical alumina nanopores with polymer melts, *Nano Lett.*, **6**(5), pp. 1075–1079.

84. Yeshchenko, O. A., Dmitruk, I. M., Koryakov, S. V., Galak, M. P. (2005). Fabrication, study of optical properties and structure of most stable $(CdP_2)_n$ nanoclusters, *Physica E*, **30**(1–2), pp. 25–30.

85. Yeshchenko, O. A., Dmitruk, I. M., Koryakov, S. V., Galak, M. P., Pundyk, I. P., Hohlova, L. M. (2005). Optical properties and structure of most stable subnanometer $(ZnAs_2)_n$ clusters, *Physica B*, **368**(1–4), pp. 8–15.

86. Yeshchenko, O. A., Dmitruk, I. M., Koryakov, S. V., Pundyk, I. P., Barnakov, Y. A. (2005). Optical study of ZnP_2 nanoparticles in zeolite Na-X, *Solid State Commun.*, **133**(2), pp. 109–112.

87. Lin, Z., Wang, Z., Chen, W., Lir, L., Li, G., Liu, Z., Han, H., Wang, Z. (1996). Absorption and Raman spectra of Se_8-ring clusters in zeolite 5A, *Solid State Commun.*, **100**(12), pp. 841–843.

88. Poborchii, V. V. (1998). Raman spectra of sulfur, selenium or tellurium clusters confined in nano-cavities of zeolite A, *Solid State Commun.*, **107**(9), pp. 513–518.

89. Kwon, S., Yan, X., Contreras, A. M., Liddle, J. A., Somorjai, G. A., Bokor, J. (2005). Fabrication of metallic nanodots in large-area arrays by mold-to-mold cross imprinting (MTMCI), *Nano Lett.*, **5**(12), pp. 2557–2562.

90. Sharpe, R. B. A., Titulaer, B. J. F., Peeters, E., Burdinski, D., Huskens, J., Zandvliet, H. J. W., Reinhoudt, D. N., Poelsema, B. (2006). Edge transfer lithography using alkanethiol inks, *Nano Lett.*, **6**(6), pp. 1235–1239.

91. Sun, S., Montague, M., Critchley, K., Chen, M., Dressick, W. J., Evans, S. D., Leggett, G. J. (2006). Fabrication of biological nanostructures by scanning near-field photolithography of chloromethylphenylsiloxane monolayers, *Nano Lett.*, **6**(1), pp. 29–33.

92. Altissimo, M., Romanato, F., Vaccari, L., Businaro, L., Cojoca, D., Kaulich, B., Cabrini, S., Fabrizio, E. D. (2002). X-ray lithography fabrication of a zone plate for X-rays in the range from 15 to 30 keV, *Microelectron. Eng.*, **61–62**(7), pp. 173–177.

93. Portavoce, A., Kammler, M., Hull, R., Reuter, M. C., Ross, F. M. (2006). Mechanism of the nanoscale localization of Ge quantum dot nucleation on focused ion beam templated Si(001) surfaces, *Nanotechnology*, **17**(17), pp. 4451–4455.

94. Juhasz, R., Linnros, J. (2002). Silicon nanofabrication by electron beam lithography and laser-assisted electrochemical size-reduction, *Microelectron. Eng.*, **61–62**(7), pp. 563–568.

95. Myers, B. D., Dravid, V. P. (2006). Variable pressure electron beam lithography (VP-eBL): a new tool for direct patterning of nanometer-scale features on substrates with low electrical conductivity, *Nano Lett.*, **6**(5), pp. 963–968.

96. Kawai, A., Suzuki, K. (2006). Removal mechanism of nano-bubble with AFM for immersion lithography, *Microelectron. Eng.*, **83**(4–9), pp. 655–658.

97. Bouchiat, V., Fauchera, M., Fournier, T., Pannetier, B., Thirion, C., Wernsdorfer, W., Clément, N. (2002). Resistless patterning of quantum

nanostructures by local anodization with an atomic force microscope, *Microelectron. Eng.*, **61–62**(7), pp. 517–522.

98. Sugimura, H., Takai, O., Nakagiri, N. (1999). Scanning probe lithography for electrode surface modification, *J. Electroanal. Chem.*, **473**(1–2), pp. 230–234.

99. Lyding, J. W., Hess, K., Abeln, G. C., Thompson, D. S., Moore, J. S., Hersam, M. C., Foley, E. T. (1998). Ultrahigh vacuum–scanning tunneling microscopy nanofabrication and hydrogen/deuterium desorption from silicon surfaces: implications for complementary metal oxide semiconductor technology, *Appl. Surf. Sci.*, **130–132**(6), pp. 221–230.

100. Zhao, J., Uosaki, K. (2002). Formation of nanopatterns of a self-assembled monolayer (SAM) within a SAM of different molecules using a current sensing atomic force microscope, *Nano Lett.*, **2**(2), pp. 137–140.

101. Hanumanthu, R., Stebe, K. J. (2006). Equilibrium shapes and locations of axisymmetric, liquid drops on conical, solid surfaces, *Colloids Surf. A*, **282–283**(7), pp. 227–239.

102. Antoncik, E. (2005). Dip-pen nanolithography: a simple diffusion model, *Surf. Sci.*, **599**(1–3), pp. pp. L369–L371.

103. Noy, A., Miller, A. E., Klare, J. E., Weeks, B. L., Woods, B. W., DeYoreo, J. J. (2002). Fabrication of luminescent nanostructures and polymer nanowires using dip-pen nanolithography, *Nano Lett.*, **2**(2), pp. 109–112.

104. Li, B., Zhang, Y., Hu, J., Li, M. (2005). Fabricating protein nanopatterns on a single DNA molecule with dip-pen nanolithography, *Ultramicroscopy*, **105**(1–4), pp. 312–315.

105. Nyamjav, D., Ivanisevic, A. (2005). Templates for DNA-templated Fe_3O_4 nanoparticles, *Biomaterials*, **26**(15), pp. 2749–2757.

106. Haaheim, J., Eby, R., Nelson, M., Fragala, J., Rosner, B., Zhang, H., Athas, G. (2005). Dip Pen Nanolithography (DPN): process and instrument performance with NanoInk's Nscriptor system, *Ultramicroscopy*, **103**(2), pp. 117–132.

107. Vijaykumar, T., John, N. S., Kulkarni, G. U. (2005). A resistless photolithography method for robust markers and electrodes, *Solid State Sci.*, **7**(12), pp. 1475–1478.

108. Bullen, D., Liu, C. (2006). Electrostatically actuated dip pen nanolithography probe arrays, *Sens. Actuators A*, **125**(2), pp. 504–511.

109. Luoh, R., Hahn, H. T. (2006). Electrospun nanocomposite fiber mats as gas sensors, *Compos. Sci. Technol.*, **66**(14), pp. 2436–2441.

110. Ding, B., Li, C., Hotta, Y., Kim, J., Kuwaki, O., Shiratori, S. (2006). Conversion of an electrospun nanofibrous cellulose acetate mat from a super-hydrophilic to super-hydrophobic surface, *Nanotechnology*, **17**(17), pp. 4332–4339.

111. Guan, H., Shao, C., Liu, Y., Yu, N., Yang, X. (2004). Fabrication of $NiCo_2O_4$ nanofibers by electrospinning, *Solid State Commun.*, **131**(2), pp. 107–109.

112. Ostermann, R., Li, D., Yin, Y., McCann, J. T., Xia, Y. (2006). V_2O_5 nanorods on TiO_2 nanofibers: a new class of hierarchical nanostructures enabled by electrospinning and calcination, *Nano Lett.*, **6**(6), pp. 1297–1302.

113. Sun, D., Chang, C., Li, S., Lin, L. (2006). Near-field electrospinning, *Nano Lett.*, **6**(4), pp. 839–842.

Chapter 3

Thermoelectric Effect of Silicon Nanowires

In this chapter, silicon single-crystal nanowires made by self-catalyzed chemical etching will be introduced first. Then, electrochemical deposition of Bi-Te nanoparticles onto the Si nanowires will be discussed. The thermoelectric properties of both silicon nanowires and Bi-Te capped Si nanowires are presented in view of the Seebeck coefficient, thermal conductivity, and the figure of merit. It is found that the Seebeck coefficient of the Si nanowire can be over 500 μV/K. The thermal conductivity could be as low as 1.7 $W \cdot m^{-1} \cdot K^{-1}$. The figure of merit is about 0.5 at 300 K. For the silicon nanowires capped with Bi-Te, the Seebeck coefficient reaches 180 μV/K. This value is four times higher than that of a single Si crystal wafer (40 μV/K).

3.1 Introduction

Thermoelectric (TE) units can convert waste heat from various sources into clean energy. Semiconducting materials are often used in such units because they have higher power factors than metallic alloys [1]. Among various semiconducting materials, silicon (Si) has been extensively studied because of the low cost and relatively high

Nanomaterials for Thermoelectric Devices
Yong X. Gan

ISBN 978-981-4774-98-7 (Hardcover), 978-0-429-48872-6 (eBook)
www.panstanford.com

figure of merit [2]. In addition, silicon has fairly good compatibility with connecting electrode materials [3]. Recently nanostructured silicon in the form of a thin film or a fine pillar for TE energy conversion has been reported [4–6]. 1D nanostructured silicon is often processed by electron-beam (E-beam) lithography [7, 8] and reduced pressure chemical vapor deposition (CVD) coupled with Au-catalyzed vapor-liquid-solid growth [9]. Metal-assisted directional chemical etching can also be used to obtain silicon nanowires (Si NWs) [10].

Semiconducting chalcogenide materials, for example, bismuth telluride compounds, have a high figure of merit in the temperature range of 200 K to 400 K [11], which allows them to be good candidates for the direct solar thermal energy conversion application [12]. However, both Bi and Te belong to the group of rare earth elements; they are expensive and very limited in supply, especially Te. One of the solutions is to manufacture composite materials to reduce the usage of these elements without sacrificing the TE energy conversion efficiency significantly. In our previous studies, Bi-Te [13], Pb-Te [14], and Bi-Te-Pb [15] were electrodeposited on nanostructured metal and/or ceramic substrates. However, the Seebeck coefficients of the chalcogenide-compound-covered materials are still very low because of the poor TE property of the substrates or support materials, including Cu, Ni, alumina, and titania, used. For example, Bi-Te nanoparticles on alumina show the absolute value of the Seebeck coefficient as low as 7 μV/K [13]. To resolve this issue, deposition of Bi-Te on a Si-NW substrate is considered by taking advantage of the relatively high Seebeck coefficient value of silicon.

This chapter provides a brief presentation on how to prepare Si-NW through metal-assisted catalytic etching. First, a simple hydrofluoric acid (HF) solution added with silver nitrate was used to etch silicon for preparing silicon single-crystal NWs. This is an atypical silver-induced self-catalysis etching process. Then, the HF solution with hydrogen peroxide was used as the etchant. Si NWs show a better TE property than those etched in HF. Finally, the Si NWs were capped with Bi-Te nanoparticles through electrochemical deposition. The TE property of the composite NWs in view of the Seebeck effect coefficient was characterized as well.

3.2 Preparation of Silicon Nanowires by Etching in a HF Solution

In this section, the main focus is on NW preparation via HF etching. First, semiconducting Si NWs were made through a chemical solution approach. Silver NWs were used as the catalyst for Si NW formation. Second, the Si NWs were decorated with Bi-Te TE material in nanoparticle form. Then the TE behavior of the Bi-Te-decorated Si NWs was investigated.

The formation of Si NWs can be described as follows. The silicon used was purchased from University Wafer with the series number of 4-100-030-V-0. It was made from single-side-polished n-type single-crystal silicon doped with As. The surface orientation is <111>. It was cleaned in a HF solution and selectively etched in a silver nitrate solution. NWs grow vertically on the surface by a top-down method. During the self-catalysis etching process, silver NWs were found first on the surface of the wafer. The removal of silicon atoms from the wafer surface was initiated underneath the silver NWs. The remaining silicon materials formed Si NWs.

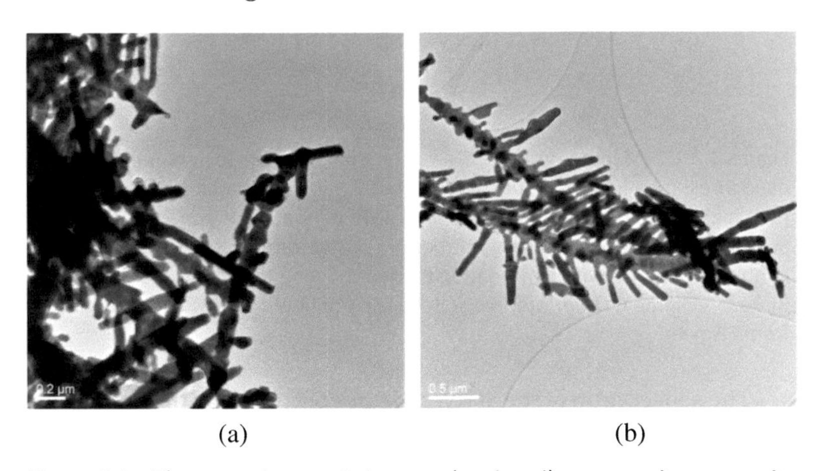

Figure 3.1 Electron microscopic images showing silver nanowires on an As-doped silicon single crystal: (a) clusters of silver nanowires and (b) highlighted fractals consisting of silver nanowires. Reprinted from Ref. [18], Copyright (2013), with permission from Elsevier.

The electron microscopic images in Fig. 3.1 show the surface morphology of the silver NWs. In Fig. 3.1a, the formation of clusters of silver NWs is illustrated. In Fig. 3.1b, the morphological details of the silver NWs in fractal configuration are revealed. More details about the growth of Si NWs were provided using the self-catalyzed chemical etching approach. First, a solution containing 50 mM silver nitrate, $AgNO_3$, was prepared. Then the solution was added into a container containing HF. The concentration of HF was 25%. The As-doped n-type Si wafer was cleaned in HF to remove the surface oxide layer and then placed into the 50 mM silver nitrate + 25% HF solution. Self-catalyzed reaction was triggered to generate the Ag and Si NWs.

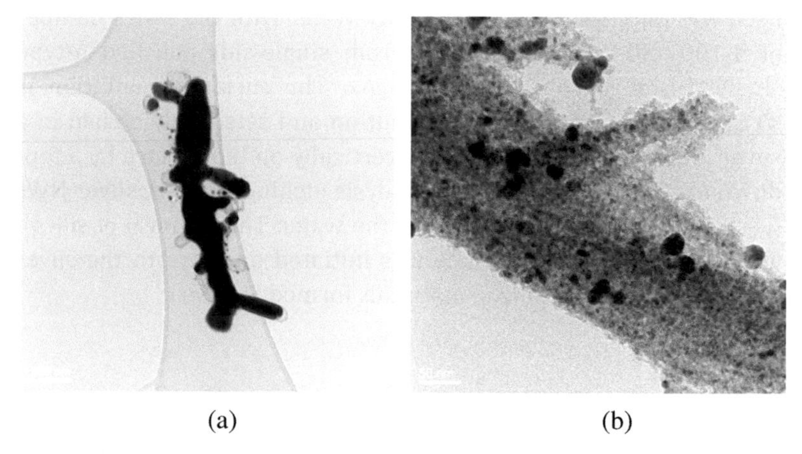

Figure 3.2 TEM images of the silver catalyst and the Bi-Te nanoparticles on Si nanowires: (a) a low-magnification image showing bifurcated silver nanowires, (b) a high-magnification image showing Bi-Te nanoparticles on etching formed Si nanowires, and (c) energy-dispersive X-ray diffraction spectrum of Si nanowires capped with Bi-Te nanoparticles showing the major elements of Si, Bi, Te, and some O. Reprinted from Ref. [18], Copyright (2013), with permission from Elsevier.

The aspect ratio of the Si NWs can be controlled by the processing parameters such as silver nitrate concentration, temperature, and reaction time, as shown in Fig. 3.2. In a fresh solution, the obtained Ag shows transition from the particle shape to the fiber form, as shown in Fig. 3.2a. It is observed that with the consumption of silver ions, the newly generated Ag product was in a small nanoparticle form. Obviously, this transition in morphology is mainly due to the availability of the silver source. This means that the concentration effect of the silver ions in the solution is obvious. The lower the silver ion concentration, the smaller is the aspect ratio of the silver nanostructure. In this work, we kept the silver ion concentration around 20 mM. If the concentration is lower than this level, it has been found that silver nanoparticles are the major products. If the silver ion concentration is higher than this value, silver is produced in dendritic NW form.

3.3 Electrodeposition of Bi-Te Nanoparticles on Silicon Nanowires

To obtain Bi-Te nanoparticles on Si NWs, a CHI 440C Electrochemical Workstation was used. The Si NWs on the silicon wafer were deposited with the Bi-Te nanoparticles. The electrodeposition process was carried out using a Pt wire as the counterelectrode and a Ag/AgCl electrode as the reference one. An electrolyte containing 0.01 M TeO_2, 0.01 M $Bi(NO_3)_2 \cdot 5H_2O$, and 1.0 M HNO_3 was used. The $Bi(NO_3)_2 \cdot 5H_2O$ was dissolved in deionized water to generate Bi^{2+} ions. TeO_2 reacted with HNO_3 to generate Te^{4+} ions. A linear potential scan with the initial, high voltage of 0.0 V was set. The lowest voltage of −0.5 V and a scan rate of 0.01 V/s were used. Figure 3.2b shows the Bi-Te nanoparticles on the Si NW fractal. To verify the chemical composition of the nanostructure, qualitative analysis of the elements using energy-dispersive X-ray (EDX) diffraction technique was performed. The results are presented in Fig. 3.2c. The EDX diffraction spectrum of the Si NWs capped with Bi-Te nanoparticles confirms the major elements of Si, Bi, Te, and O.

Table 3.1 lists the atomic percentage of each element. Quantitative analysis indicates that the atomic ratio of Bi to Te is about 1:1. The Bi-Te nanoparticles are not just physical admixtures. As previously reported using the similar nitric acid–based electrolyte, the electrically codeposited Bi and Te nanoparticles tend to form a stoichiometric compound, Bi_2Te_3 [16, 17]. But from our early study [13], it was found that the weight percent of Bi:Te is 62%:38%. This value corresponds to the atomic ratio of 1:1, which is exactly the same result as we obtained in this study. Therefore, it is believed that the obtained Bi-Te nanoparticles in this work are in the compound form of Bi-Te. Its composition is slightly shifted away from the stoichiometric composition of Bi_2Te_3. Since there is excessive Te in

Bi-Te obtained in this work, this Bi-Te compound should be an n-type TE material.

Table 3.1 EDX elemental analysis results [18]

Element	Cu	Ag	Si	Te	Bi	O
Atomic%	24.37	1.420	44.00	15.23	14.95	0.0400

3.4 Seebeck Coefficient of Silicon Nanowires

The Seebeck coefficient of the Si/Bi-Te NW was measured using a self-build measurement system containing a Talboys heat platform with temperature control and a mode 410 Extech Multimeter. The NWs were removed from the wafer and bonded by a silver adhesive. Then they were set on an alumina plate. A temperature difference was imposed at the two ends of the specimen. The average absolute values of the Seebeck coefficients obtained at different measuring temperature ranges were plotted and shown in Fig. 3.3a. It is found that the average Seebeck value for the NW is 180 μV/K. As compared with the silicon bulk material, the NW has a Seebeck value that is four times higher. The measured Seebeck value of the silicon bulk material is only about 40 μV/K under the same test conditions.

Also measured is the Seebeck coefficient of the Si NW without any Bi-Te nanoparticle. The results are shown in Fig. 3.3b. The average value of the measurement is 70 μV/K, with a standard derivation of 8 μV/K. As compared with the results shown in Fig. 3.3a, the Bi-Te nanoparticle–capped Si NWs show a higher Seebeck coefficient. It is evident that the TE property of the Si NW is enhanced significantly on adding Bi-Te nanoparticles to form a composite nanomaterial.

The increase in the Seebeck coefficient of the silicon/Bi-Te nanomaterial may be explained by the quantum confinement in the Bi-Te nanoparticle/Si NW structure [18]. The Si NW fractals show a stronger Seebeck effect than the bulk silicon. The Bi-Te nanoparticles are separated from each other on the surface of the Si NW. This could further enhance the Seebeck coefficient because of the reduced electron mobility from the isolated chacolgenide compound.

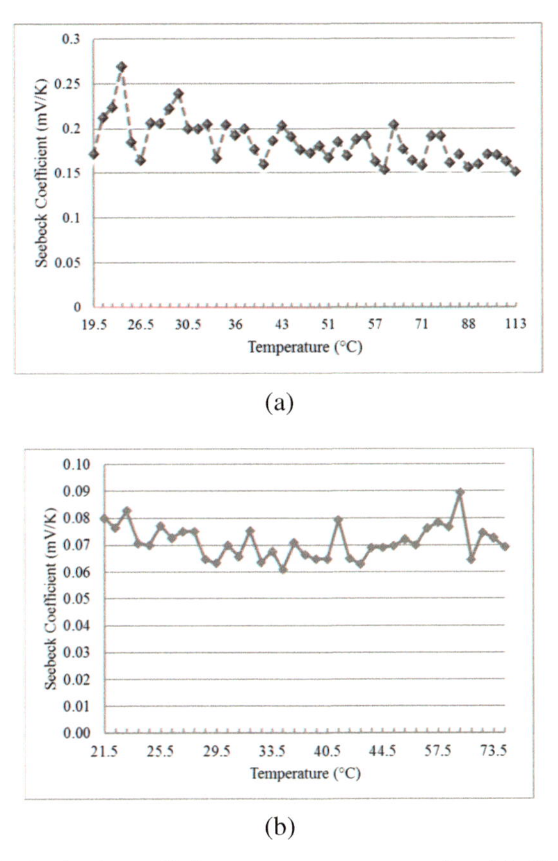

Figure 3.3 Seebeck coefficient measurement results: (a) Si nanowires decorated with Bi-Te nanoparticles and (b) Si nanowires without Bi-Te nanoparticles. Reprinted from Ref. [18], Copyright (2013), with permission from Elsevier.

3.5 Silicon Nanowire Arrays with an Improved Thermoelectric Property

In the work performed by Zhang et al. [19], high-density and large-area vertically aligned porous Si NW arrays (Si-NWAs) with different morphologies, such as various lengths, porosities, and heterogeneous diameters, were fabricated on the two sides of the silicon substrate using the silver-assisted chemical etching method.

The silver nanoparticles were deposited on silicon in a solution of 0.01 M $AgNO_3$ and 4.8 M HF for 1 min. The silver nanoparticle–covered silicon wafer was immersed in a solution of 0.2 M H_2O_2 and 4.8 M HF for etching. Due to the double-sided etching, Si NW arrays developed on both sides to form a sandwich structured composite (SSC), as shown in Fig. 3.4. The Seebeck coefficient of the Si NW measured was over 500 μV/K. The thermal conductivity could be as low as 1.7 $W \cdot m^{-1} \cdot K^{-1}$. The figure of merit of this SSC was calculated to be as high as 0.5 at 300 K. The size and morphology of the Si NWs are controlled by the etching time.

Figure 3.4 Illustration and cross-section view of silicon nanowire arrays in the form of a sandwich structured composite (SSC). Reprinted from Ref. [19], Copyright (2015), with permission from Elsevier.

In view of how thermal conductivity can be reduced, Feser et al. [20] did fundamental studies and found that the thermal conductivity of Si NW arrays can be changed by controlling the pattern and roughness of the NWs. Early work [21] has shown that using a combined nanoprinting and metal-assisted chemical etching manufacturing process, Si NW arrays with a defined geometry can be made.

3.6 Conclusions

In summary, silicon single-crystal NWs can be successfully made by self-catalyzed chemical etching. Bi-Te nanoparticles are deposited onto the NWs via electroplating. The Bi-Te-capped NWs show enhanced TE behavior. The absolute value of the Seebeck coefficient of the Si NW capped with Bi-Te, 180 μV/K, is four times higher than that of the single Si crystal wafer, which is 40 μV/K. The Seebeck coefficient of the Si NW capped with Bi-Te is two times higher than that of the single Si crystal NW, 70 μV/K, on etching with HF. It is obvious that the TE property of the Si NW is better than that of the silicon bulky material (the silicon wafer). The Seebeck effect of the Si NW can be enhanced significantly by adding Bi-Te nanoparticles on it to form a composite nanomaterial. The increase in the Seebeck coefficient of the silicon/Bi-Te NW is due to the quantum confinement in the nanostructure. Using a mixture of HF and hydrogen peroxide etchant, the Seebeck coefficient of the Si NW obtained reached 500 μV/K. The thermal conductivity of the Si NW SSC can be reduced to 1.7 $W \cdot m^{-1} \cdot K^{-1}$. The figure of merit of this SSC was about 0.5 at 300 K by calculation.

References

1. Gonçalves, A. P., Lopes, E. B., Delaizir, G., Vaney, J. B., Lenoir, B., Piarristeguy, A., Pradel, A., Monnier, J., Ochin, P., Godart, C. (2012). Semiconducting glasses: a new class of thermoelectric materials? *J. Solid State Chem.*, **193**(9), pp. 26–30.
2. Ci, P., Shi, J., Wang, F., Xu, S., Yang, Z., Yang, P., Wang, L., Chu, P. K. (2011). Novel thermoelectric materials based on boron-doped silicon microchannel plates, *Mater. Lett.*, **65**(11), pp. 1618–1620.
3. Gan, Y. X., Dynys, F. W. (2013). Joining highly conductive and oxidation resistant silver-based electrode materials to silicon for high temperature thermoelectric energy conversions, *Mater. Chem. Phys.*, **138**(1), pp. 342–349.
4. Dávila, D., Tarancón, A., Calaza, C., Salleras, M., Fernández-Regúlez, M., San Paulo, A., Fonseca, L. (2012). Monolithically integrated thermoelectric energy harvester based on silicon nanowire arrays for powering micro/nanodevices, *Nano Energy*, **1**(6), pp. 812–819.
5. Chen, Z. G., Han, G., Yang, L., Cheng, L., Zou, J. (2012). Nanostructured thermoelectric materials: current research and future challenge, *Prog. Nat. Sci.*, **22**(6), pp. 535–549.
6. Stranz, A., Sökmen, Ü., Kähler, J., Waag, A., Peiner, E. (2011). Measurements of thermoelectric properties of silicon pillars, *Sens. Actuators A*, **171**(1), pp. 48–53.

7. Wu, J. Y., Tseng, C. L., Wang, Y. K., Yu, Y., Ou, K. L., Wu, C. C. (2013). Detecting Interleukin-1β genes using a N_2O plasma modified silicon nanowire biosensor, *J. Exp. Clin. Med.*, **5**(1), pp. 12–16.
8. Wessely, F., Krauss, T., Schwalke, U. (2013). CMOS without doping: multi-gate silicon-nanowire field-effect-transistors, *Solid-State Electron.*, **70**(4), pp. 33–38.
9. Serre, P., Ternon, C., Stambouli, V., Periwal, P., Baron, T. (2013). Fabrication of silicon nanowire networks for biological sensing, *Sens. Actuators B*, **182**(6), pp. 390–395.
10. Li, X. (2012). Metal assisted chemical etching for high aspect ratio nanostructures: a review of characteristics and applications in photovoltaics, *Curr. Opin. Solid State Mater. Sci.*, **16**(2), pp. 71–81.
11. Venkatasubramanian, R., Siivola, E., Colpitts, T., O'Quinn, B. (2001). Thin-film thermoelectric devices with high room-temperature figures of merit, *Nature*, **413**(6856), pp. 597–602.
12. Tritt, T. M., Funahashi, R., Boettner, H. (2015). Advances in thermoelectric materials II introduction, *J. Mater. Res.*, **30**(17), pp. 2517–2517.
13. Gan, Y. X., Sweetman, J., Lawrence, J. G. (2010). Electrodeposition and morphology analysis of Bi–Te thermoelectric alloy nanoparticles on copper substrate, *Mater. Lett.*, **64**(3), pp. 449–452.
14. Madhavaram, R., Sander, J., Gan, Y. X., Masiulaniec, C. K. (2009). Thermoelectric property of PbTe coating on copper and nickel, *Mater. Chem. Phys.*, **118**(1) pp. 165–173.
15. Su, L., Gan, Y. X. (2011). Formation and thermoelectric property of TiO_2 nanotubes covered by Te–Bi–Pb nanoparticles, *Electrochim. Acta*, **56**(16), pp. 5794–5803.
16. Sander, M. S., Gronsky, R., Sands, T., Stacy, A. M. (2003). Structure of bismuth telluride nanowire arrays fabricated by electrodeposition into porous anodic alumina templates, *Chem. Mater.*, **15**(1), pp. 335–339.
17. Golia, S., Arora, M., Sharma, R. K., Rastogi, A. C. (2003). Electrochemically deposited bismuth telluride thin films, *Curr. Appl. Phys.*, **3**(2–3), pp. 195–197.
18. Gan, Y. X., Koludrovich, M. J., Zhang, L. (2013). Thermoelectric effect of silicon nanofibers capped with Bi–Te nanoparticles, *Mater. Lett.*, **111**(11), pp. 126–129.

19. Zhang, T., Wu, S., Xu, J., Zheng, R., Cheng, G. (2015). High thermoelectric figure-of-merits from large-area porous silicon nanowire arrays, *Nano Energy*, **13**(4), pp. 433–441

20. Feser, J. P., Sadhu, J. S., Azeredo, B. P., Hsu, K. H., Ma, J., Kim, J., Seong, M., Fang, N. X., Li, X., Ferreira, P. M., Sinha, S., Cahill, D. G. (2012). Thermal conductivity of silicon nanowire arrays with controlled roughness, *J. Appl. Phys.*, **112**, Article No. 114306.

21. Chern, W., Hsu, K., Chun, I. S., de Azeredo, B. P., Ahmed, N., Kim, K. H., Zuo, J., Fang, N., Ferreira, P., Li, X. (2010). Nonlithographic patterning and metal-assisted chemical etching for manufacturing of tunable light-emitting silicon nanowire arrays, *Nano Lett.*, **10**(5), pp. 1582–1588.

Chapter 4

Electrodeposited Te-Bi-Pb Thermoelectric Films

This chapter presents electrochemical de-alloying of zinc from C26000 brass ($Cu_{70}Zn_{30}$ alloy) to form porous copper, followed by electrodeposition of Te-Bi-Pb alloys on the porous substrate. Te-Bi-Pb alloys are deposited in a nonequilibrium condition with linearly increasing cathode potentials. Te has the highest tendency to get deposited on the substrate, while Bi and Pb have the same tendency. Scanning electron microscopic analysis indicates that with an increase in the thickness of the Te-Bi-Pb coating, the morphology of the alloys changes from particulate to dendritic fractals. The composition of the alloys changes from Te_2BiPb to Te_3BiPb. The thermoelectricity of the deposited alloys on porous copper was tested. The results show that Te-Bi-Pb is an n-type semiconductor. Due to the deposition of Te-Bi-Pb alloys on the copper surface, the absolute value of the Seebeck coefficient increases.

4.1 Introduction

Thermoelectric (TE) materials can capture waste heat and convert it into electrical energy. Researchers are considering building biothermal batteries, cooling devices, and power generation units with TE materials. Traditional semiconducting materials, like

Nanomaterials for Thermoelectric Devices
Yong X. Gan

ISBN 978-981-4774-98-7 (Hardcover), 978-0-429-48872-6 (eBook)
www.panstanford.com

amorphous silicon [1], may not have sufficient energy conversion efficiency. Therefore, there is a need for developing various new TE materials through different material processing techniques. Molecular beam epitaxy (MBE) is a powerful method for growing thin films. This method has been used for making p-Si/$Si_{0.5}Ge_{0.5}$ superlattices (SLs) on a p-type silicon single crystalline substrate. Due to the TE effect, transient laser-induced voltage (LIV) signals were generated by the SL materials [2, 3]. The radio frequency (RF) plasma processing technique has been used to prepare Al-doped ZnO nanopowders [4, 5]. The powders were hot-pressed to form pellets. Specimens made of the pellets showed good TE property and high electrical conductivity comparable to those of metals. Porosity, which can be determined by resonant ultrasound spectroscopy, affects the TE properties of materials [6]. Spark plasma–sintered high-density polycrystalline $CaCd_2Sb_2$ showed good TE properties [7]. Nearly single phased TiNiSn metallic compound TE materials were synthesized by combining mechanical alloying (MA) and spark plasma sintering (SPS) [8, 9]. The thermal conductivity of TiNiSn can be reduced by refining the grain sizes, which is good for increasing the TE energy conversion efficiency. The vacuum melting method was used for synthesis of p-type $Zn_4Sb_{(3-x)}Te_x$ (x = 0.02–0.08). The electrical conductivity and Seebeck coefficient in the temperature range of 300–700 K were calculated [10]. It was found that the TE figure of merit (*ZT*) was increased with the increase of Te content. Bi_2Se_3 and Bi_2O_3 were synthesized by solid-state reaction in an evacuated quartz ampoule to calculate the dimensionless *ZT* [11]. The *ZT* value was about 0.2 at 800 K [11, 12]. TE properties are temperature sensitive [13]. This is because the conductivity of materials varies with temperature. For example, $ReGa_xSe_yTe_{(15-y)}$ ($0 \leq x \leq 2$; $0 \leq y \leq 7.5$) was studied in the temperature range of 90–320 K [14]. There exists hopping conduction in the temperature range of 150 K to 280 K for all samples.

Electrodeposition has been used for obtaining thin films with many new properties. It has also been considered for preparing nanostructures, for example, fabricating tin oxide nanowires (NWs) through one-step deposition [15] and synthesizing highly active Ag/Pd nanorings for activating electroless copper deposition [16]. Zhou et al. [17] presented a method to obtain electrodeposited NiFeMo thin films with surface roughness and good magnetic properties.

Song [18] proposed and developed a technique for rapid-coating magnesium alloys by utilizing the high surface alkalinity of the alloys. The coating is highly stable and can offer good protection under service conditions. Improvement in the stability and hardness of porous silicon structures by electrochemical deposition of silicon capping was also reported [19]. Deposition of Cu-CoWP/NiB composite films was investigated, and the relation between the element content and the electrical resistivity of the deposit was established [20]. For the last couple of decades, researchers have made intensive studies on electrodeposition of TE materials. For example, bismuth telluride thin films with high thermopowers were obtained using this approach [21]. The structure, morphology, and TE property of the electrodeposited Bi-Te were studied [22]. It is also reported that Pb-Te alloys can be electrodeposited on copper and nanoporous nickel [23].

The objective of this chapter is to show how to prepare Te-Bi-Pb alloys on a porous substrate. First, we will introduce the electrochemical de-alloying of zinc from C26000 brass ($Cu_{70}Zn_{30}$ alloy). How to obtain a porous copper substrate will be discussed. Then, electrodeposition of Te-Bi-Pb on the porous substrate will be detailed. The deposition of Te-Bi-Pb alloys occurred in a nonequilibrium condition with linearly increasing cathode potentials. Scanning electron microscopic (SEM) analysis was performed to reveal the morphology of Te-Bi-Pb. The composition of the alloy was analyzed by energy-dispersive X-ray spectroscopy (EDS). Preliminary tests on thermoelectricity of the deposited alloys on porous copper were also conducted.

4.2 Experimental Methods

A 1.0 M $CuSO_4$ solution was made by mixing 4.8 g of $CuSO_4 \cdot 5H_2O$ with water. This solution is used as an electrolyte to create nonporous copper. Three strips of brass were cut from a sheet to a width of about 2 mm and were cleaned with acetone. The brass strips were then placed in a copper sulfate pentahydrate solution with two electrodes; the reference electrode was Ag/AgCl, and the other was a platinum wire counterelectrode. The brass foil was made of Cu:Zn 70:30 wt%. Using the CHI400A programmable electrochemical workstation, we were able to run cyclic voltammetry tests. The first positive scan to remove both zinc and copper had the following parameters: high potential, or high E(v), = 0.5 V; low potential, or low E(v), = 0 V; and sensitivity 1×10^{-2} V/s. The negative scan to deposit nanoporous copper (NPCu) ran with the following parameters: high E(v) = +0.5 V and low E(v) = −1.0 V, with the sensitivity left the same. A higher positive potential was chosen in the second scanning cycle. The second positive scan ran with the high voltage of 1.0 V.

After the NPCu was prepared, a TE layer was coated onto the porous substrate. The solution used for electrodepositing Te-Bi-Pb contained 0.05 M $Pb(OOCCH_3)_2 \cdot 3H_2O$, 0.01 M TeO_2, 0.01 M $Bi(NO_3)_3 \cdot 5H_2O$, and 1.0 M HNO_3. To make the electrolyte, first, 0.03 g of TeO_2 was dissolved into 1.8 mL of 70% HNO_3 at 80°C. Then, 0.10 g of $Bi(NO_3) \cdot 5H_2O$ and 0.40 g of $Pb(OOCCH_3)_2 \cdot 3H_2O$ were added. Finally, to have the required concentrations, the solution was diluted by adding water so that the total volume of the solution was 20 mL. After that, Te-Bi-Pb was electrodeposited onto the surface, creating a thermoelectric coating (TEC). SEM micrographs were taken, and the composition of Te-Bi-Pb was analyzed using an Oxford Instruments EDX detector mounted on the Hitachi S4800 scanning electron microscope. The Seebeck effect for NPCu with and without the TEC was tested. The specimens under testing were put into two beakers containing 1.0 M KCl (1.15 g of KCl mixed with 15 mL of water). The first beaker was kept at room temperature (25°C); the second beaker was heated to 80°C. The electric potential difference of the materials was tested. Four tests were conducted on the base metal and the copper with the TEC. The results in the form of open circuit potential data were recorded for analysis.

4.3 Results and Discussion

The SEM image in Fig. 4.1a shows the microstructure of Te-Bi-Pb on the NPCu substrate. The dark region is the uncovered base material of NPCu. The white portion stands for the TE material, Te-Bi-Pb alloy. From the morphology of Fig. 4.1a, it can be seen that the Te-Bi-Pb alloy forms clusters consisting of many nanoparticles. The size of the clusters is in the range of several microns. Figure 4.1b shows the elemental analysis results of Te-Bi-Pb on the copper alloy substrate

(a)

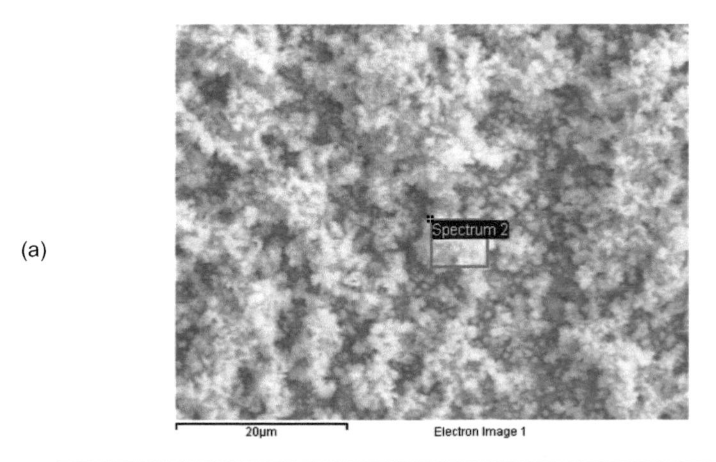

(b)

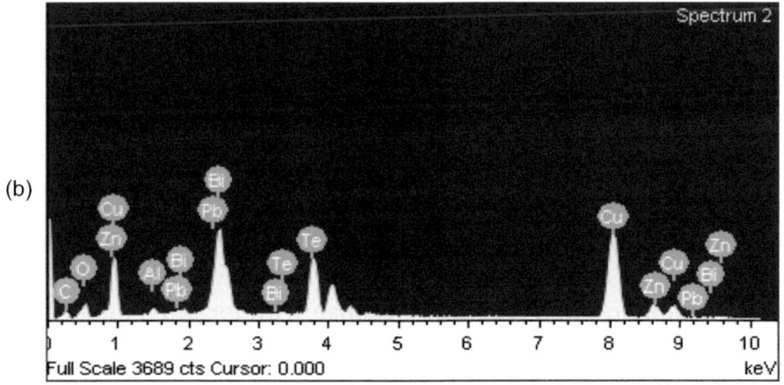

(c)

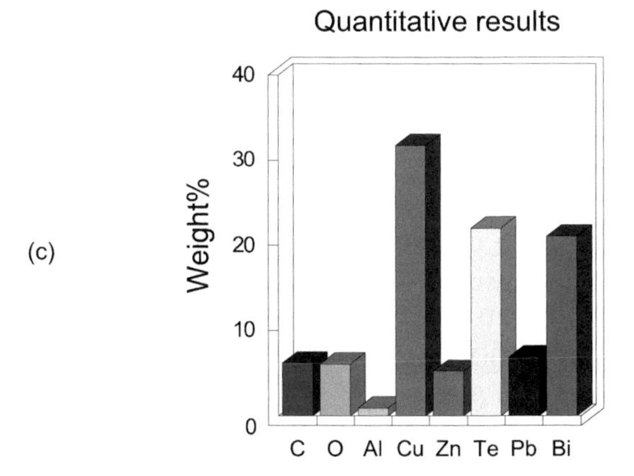

Figure 4.1 SEM analysis of Te-Bi-Pb on a substrate: (a) morphology, (b) EDX spectrum, and (c) composition.

using EDS. The composition is Te, Bi, Pb, Cu, and Zn. Signals from the copper substrate were detected, as shown by the three Cu peaks and three Zn peaks, because the TE material does not cover the substrate completely. Small peaks corresponding to Al, O, and C as the impurities in the materials were also found. Oxygen is believed to be the result of surface oxidation of copper and zinc. For Te, the highest peak is at 3.8 keV. For Bi, the highest peak is located at 2.2 keV. For Pb, the highest peak is at 2.45 keV. Quantitative analysis results shown in Fig. 4.1c indicate that the ratio in weight percent of Te:Bi:Pb = 24%:20%:8% (or 6:5:2), which corresponds to the atomic ratio of Te:Bi:Pb = 2:1:1. This atomic ratio corresponds to the formula of Te_2BiPb.

We also analyzed the structure and composition of the Te-Bi-Pb material located well above the substrate. The morphology shows dendrite form or fractals of the Te-Bi-Pb alloy as illustrated by the SEM image in Fig. 4.2a. The EDS results taken from the Te-Bi-Pb fractals are shown in Fig. 4.2b. On examining the peaks in Fig. 4.2b, a trivial amount of Al and Tc was found. Copper, oxygen, and zinc peaks were also revealed. Since lead and bismuth are two elements with very similar atomic weights, their main peaks are close to each other (located at 2.2 keV for Bi and at 2.45 keV for Pb).

To examine the composition change with location, quantitative analysis results are shown in Fig. 4.2c. The ratio in weight percent of Te:Bi:Pb is 27%:14.5%:12.5%. Converting this weight percentage ratio into atomic ratio, we found that Te:Bi:Pb = 3:1:1. This atomic ratio corresponds to the formula of Te_3BiPb. Obviously, the content of Te in the dendrites or fractal structure of Te-Bi-Pb is significantly higher than that in the initially deposited clusters on the substrate.

Electrochemical analysis was carried out to reveal the deposition mechanism of Te-Bi-Pb. First, an experiment was set up to perform the electrodeposition of the individual elements within the TEC. The first part of the test consisted of a solution of 18 mL of water, 2 mL (1.26 g) of HNO_3, and 0.4 g of Pb salt. This solution has the concentration of 0.05 M $Pb(OOCCH_3)_2 \cdot 3H_2O$ and 1.0 M HNO_3. Using the cyclic voltammography (CV) scanning device, a typical CV curve associated with the Pb deposition was recorded, as shown in Fig. 4.3a. In the negative scan region, that is the cathodic polarization stage, the lead was drawn out of the solution and got deposited onto the nanoporous electrode. During the forward scan, Pb was

(a)

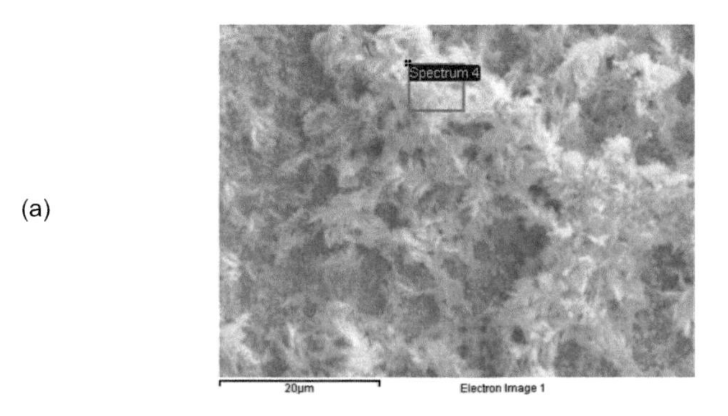

(b)

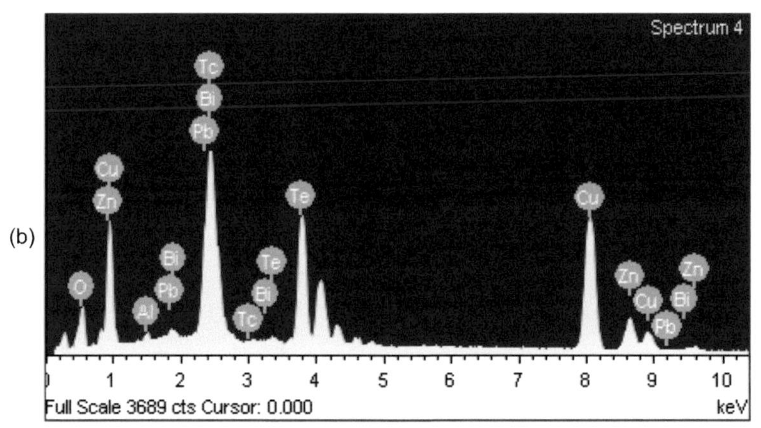

(c)

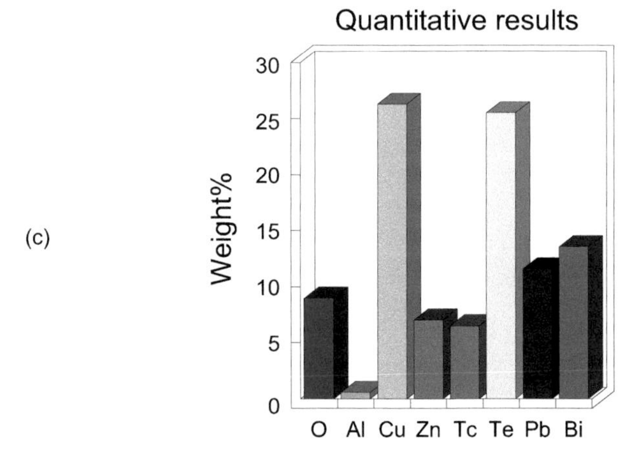

Figure 4.2 SEM analysis of Te-Bi-Pb fractals: (a) morphology, (b) EDX spectrum, and (c) composition.

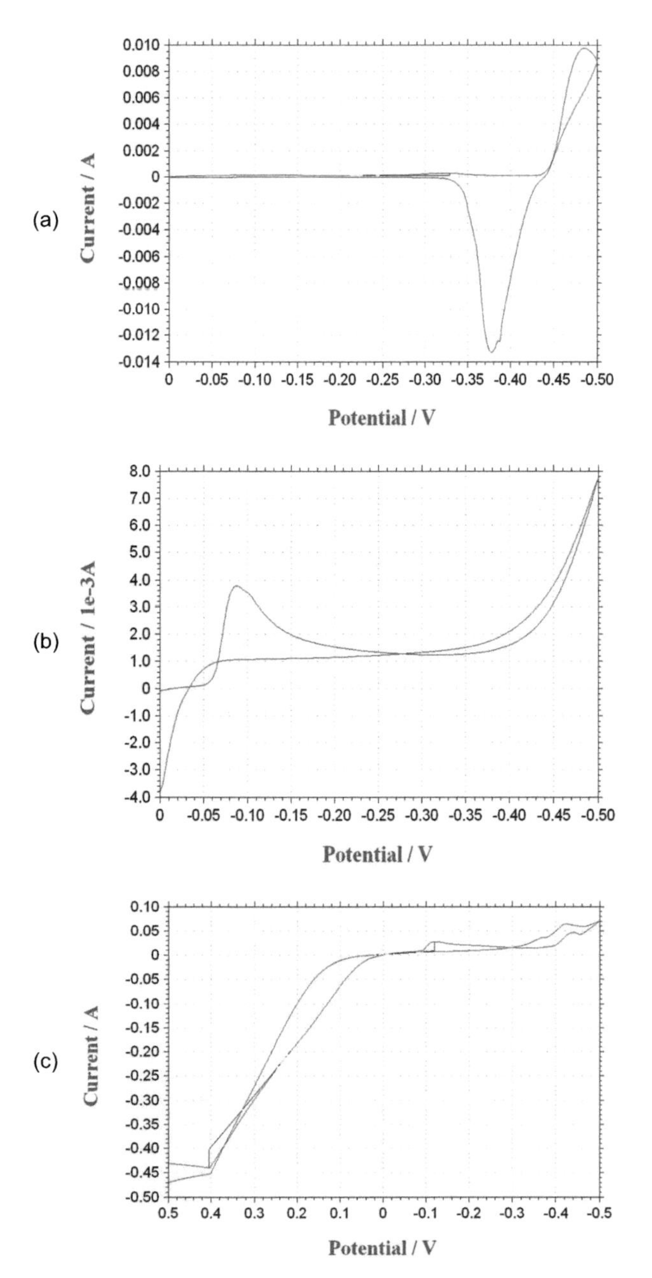

Figure 4.3 Cyclic voltammograms of electrodeposition and de-alloying: (a) Pb deposition, (b) Bi deposition, and (c) Zn de-alloying from brass to form a nanoporous copper alloy followed by Te-Bi-Pb deposition.

electrochemically deposited on the NPCu at around −0.45 V to −0.475 V, as indicated by the current peak shown in Fig. 4.3a. After the test of Pb deposition, the deposition of Bi was tested. Again, 18 mL of water was used with 1.26 g of HNO_3 and 0.1 g of Bi to make an electrolyte containing 0.01 M $Bi(NO_3)_3 \cdot 5H_2O$ and 1.0 M HNO_3. The same CV scanning process was repeated for Bi as was done for Pb. Figure 4.3b shows the Bi deposition peak at around −0.06 V to −0.10 V, revealing that Bi is easier to deposit than Pb onto the NPCu alloy. As can be seen in Fig. 4.3c, Te and Pb have very similar deposition potentials. The peak at around −0.38 V refers to the deposition of Te, which is very close to the deposition potential of Pb. The CV results of de-alloying of Zn from the Cu-Zn alloy and the deposition Te-Bi-Pb in one step were plotted in the same figure. As shown in Fig. 4.3c, zinc was drawn out of the brass alloy and went into the solution in the positive scanning region (in the anodic polarization cycle).

Finally, we did some preliminary studies on the TE behavior of NPCu and the deposited Te-Bi-Pb alloy on the substrate. The potential difference between NPCu with and without the TEC was compared. Four tests for each metal sample were done under the same parameters (time 200 s and temperature difference 55°C). The Seebeck coefficient (the potential difference divided by the temperature difference) for the NPCu has an average value of about −6.27 μV/K. The Seebeck coefficient obtained from the tests for the Te-Bi-Pb coating on the NPCu is −6.7 μV/K. Since the Seebeck coefficient for the Te-Bi-Pb coating on the NPCu takes a negative value, the Te-Bi-Pb coating is an n-type semiconducting TE material.

4.4 Conclusions

Electrochemical de-alloying has resulted in the removal of zinc from C26000 brass ($Cu_{70}Zn_{30}$ alloy) to form porous copper. The Te-Bi-Pb alloy can be electrochemically deposited on the porous substrate in a nonequilibrium condition with linearly increasing cathode potentials. During the electrodeposition process, Te has the highest tendency to get deposited on the substrate, while Bi and Pb have the same tendency. SEM analysis indicates that with the increase in the thickness of the Te-Bi-Pb coating, the morphology of the alloys changes from particulate to dendrite fractals. The composition of the alloys also changes from Te_2BiPb to Te_3BiPb. Thermoelectricity tests on the deposited alloys show that the Te-Bi-Pb materials are n-type semiconductors. Deposition of Te-Bi-Pb alloys on the copper surface increases the absolute value of the Seebeck coefficient.

References

1. Goesmann, F., Jones, D. I. (1994). Thermoelectric-power of amorphous-silicon under illumination, *Philos. Mag. B*, **69**(2), pp. 159–167.
2. Zhao, S. Q., Liu, W. W., Yang, L. M., Zhao, K., Liu, H., Zhou, N., Wang, A. J., Zhou, Y. L., Zhou, Q. L., Shi, Y. L. (2009). Lateral photovoltage of B-doped ZnO thin films induced by 10.6 mu m CO_2 laser, *J. Phys. D: Appl. Phys.*, **42**(18), Article No. 185101.
3. Zou, H. L., Row, D. M., Min, G. (2001). Growth of p- and n-type bismuth telluride thin films by co-evaporation, *J. Cryst. Growth*, **222**(1–2), pp. 82–87.
4. Cheng, H., Xu, X. J., Hng, H. H., Ma, J. (2009). Characterization of Al-doped ZnO thermoelectric materials prepared by RF plasma powder processing and hot press sintering, *Ceram. Int.* **35**(8), pp. 3067–3072.
5. Ji, X. H. (2005). Syntheses and properties of nanostructured Bi_2Te_3-based thermoelectric materials, PhD dissertation, Zhejiang University, Hangzhou, China.
6. Ni, J. E., Ren, F., Case, E. D., Timm, E. J. (2009). Porosity dependence of elastic moduli in LAST (Lead–antimony–silver–tellurium) thermoelectric materials, *Mater. Chem. Phys.*, **118**(2–3), pp. 459–466.
7. Zhang, H., Fang, L., Tang, M. B., Chen, H. H., Yang, X. X., Guo, X., Zhao, J. T., Grin, Y. (2010). Synthesis and properties of $CaCd_2Sb_2$ and $EuCd_2Sb_2$, *Intermetallics*, **18**(1), pp. 193–198.
8. Zou, M., Li, J. F., Du, B., Liu, D., Kita, T. (2009). Fabrication and thermoelectric properties of fine-grained TiNiSn compounds, *J. Solid State Chem.*, **182**(11), pp. 3138–3142.
9. Ji, X. H., Zhao, X. B., Zhang, Y. H., Lu, B. H., Ni, H. L. (2005). Synthesis and properties of rare earth containing Bi_2Te_3 based thermoelectric alloys, *J. Alloys Compd.*, **387**(1–2), pp. 282–286.
10. Li, W., Zhou, L., Li, Y., Jiang, J., Xu, G. (2009). Effect of yttrium addition and the powder size on oxide film and powder ignition temperature of AZ91D magnesium alloy, *J. Alloys Compd.*, **481**(1–2), pp. 373–378.

11. Ruleova, P., Drasar, C., Lostak, P., Li, C. P., Ballikaya, S., Uher, C. (2010). Thermoelectric properties of Bi_2O_2Se, *Mater. Chem. Phys.*, **119**(1–2), pp. 299–302.

12. Dresselhaus, M., Dresselhaus, G., Avouris, P. (2001). *Carbon Nanotubes: Synthesis, Structure, Properties, and Applications*, Berlin: Springer-Verlag Berlin Heidelberg, pp. 70–78

13. Wang, L., Cai, K., Wang, Y., Li, H., Wang, H. (2009). Thermoelectric properties of indium-filled skutterudites prepared by combining solvothermal synthesis and melting, *Appl. Phys. A*, **97**(4), pp. 841–845.

14. Dalafave, D. S. (2010). Thermoelectric properties of $Re_6Ga_xSe_yTe_{15-y}$ ($0 \leq x \leq 2$; $0 \leq y \leq 7.5$), *Mater. Chem. Phys.*, **119**(1–2), pp. 195–200.

15. Ishizaki, T., Saito, N., Takai, O. (2009). Surfactant-assisted fabrication of tin oxide nanowires through one-step electrochemically induced chemical deposition, *J. Electrochem. Soc.*, **156**(10), pp. D413–D417.

16. Lee, C. L., Tseng, C. M., Wu, R. B., Syu, S. C. (2009). Synthesis of highly active Ag/Pd nanorings for activating electroless copper deposition, *J. Electrochem. Soc.*, **156**(9), pp. D348–D350.

17. Zhou, Q., Velleuer, J., Heard, P. J., Schwarzacher, W. (2009). Surface roughness and magnetic properties of electrodeposited NiFeMo thin films, *Electrochem. Solid-State Lett.*, **12**(3), pp. D7–D10.

18. Song, G. L. (2009). 'Electroless' E-coating: en innovative surface treatment for magnesium alloys, *Electrochem. Solid-State Lett.*, **12**(10), pp. D77–D79.

19. Ali, N. K., Hashim, M. R., Aziz, A. A. (2009). Pulse current electrochemical deposition of silicon for porous silicon capping to improve hardness and stability, *Electrochem. Solid-State Lett.*, **12**(3), pp. D11–D14.

20. Osaka, T., Aramaki, H., Yoshino, M., Ueno, K., Matsuda, I., Shacham-Diamand, Y. (2009). Fabrication of electroless CoWP/NiB diffusion barrier layer on SiO_2 for ULSI devices, *J. Electrochem. Soc.*, **156**(9), pp. H707–H710.

21. Golia, S., Arora, M., Sharma, R. K., Rastogi, A. C. (2003). Electrochemically deposited bismuth telluride thin films, *Curr. Appl. Phys.*, **3**(2–3), pp. 195–197.

22. Gan, Y. X., Sweetman, J., Lawrence, J. G. (2010). Electrodeposition and morphology analysis of Bi–Te thermoelectric alloy nanoparticles on copper substrate, *Mater. Lett.*, **64**(3), pp. 449–452.

23. Madhavaram, R., Sander, J., Gan, Y. X., Masiulaniec, C. K. (2009). Thermoelectric property of PbTe coating on copper and nickel, *Mater. Chem. Phys.*, **118**(1), pp. 165–173.

Chapter 5

Conducting Polymer Thermoelectric Composite Materials

In this chapter, an overview of conducting polymer materials and their composite materials for thermoelectric energy conversion is provided. Typical conducting polymers and the related processing methodologies are introduced. Following that, we present a case study on the thermoelectric property of a polyaniline (PANI)-coated TiO_2 nanotube (NT) nanocomposite. The conducting polymer, PANI, was prepared by electrochemical oxidation on the NTs. The morphology of the nanocomposites was observed. The highest absolute value of the Seebeck coefficient for the TiO_2 NT/polyaniline nanocomposite materials was found to be 124 μV/K.

5.1 Introduction

Polymeric thermoelectric (TE) materials possess some advantages over their metallic or ceramic counterparts. For example, polymeric materials have much lower densities and much lower thermal conductivity than metals or ceramics. They are flexible due to the varying molecular structures. Several families of conducting polymers, including poly(3,4-ethylenedioxythiophene) (PEDOT), polyaniline (PANI), polythiophene (PTh), polypyrrole (PPy), polyacetylene (PA), polycarbazole (PC), and their composites, have

Nanomaterials for Thermoelectric Devices
Yong X. Gan

ISBN 978-981-4774-98-7 (Hardcover), 978-0-429-48872-6 (eBook)
www.panstanford.com

been studied for TE energy conversion applications, as reviewed by Gao and Chen [1]. Conducting polymers may be mixed with inorganic nanoparticles to form hybrid composite materials. For example, in the work performed by Jundale et al. [2], chemically synthesized PANI with 10–50 wt% of CuO nanoparticles was spin-coated to form hybrid composite thin films. The thickness of the films is in the range of 220 to 340 nm. CuO nanoparticles increase the resistivity of the films. The thermopower measurement shows that PANI has a p-type behavior, while the CuO nanoparticle is n-type. If the substrate is not a conducting polymer, the electrical conductivity of the composite material may be increased by adding a metal-based TE component. For example, Plochmann et al. [3] studied a polysiloxane polymer matrix composite film filled with micrometer-scale TE $Sn_{0.85}Sb_{0.15}O_2$ particles. It was found that the electrical conductivity of the insulating polymer increased by orders of magnitude due to the percolation effect of the metallic filler.

The screen-printing process has been used to make an inorganic TE thick film and organic conducting polymer hybrid composite [4]. For making an n-type TE flexible film, a slurry consisting of Bi and Te powders and an organic binder was screen-printed on a polyimide substrate following by annealing at 450°C in a nitrogen atmosphere. Similarly, a p-type TE flexible film was made by screen-printing a slurry containing Sb and Te powders. After high-temperature annealing, the screen-printed TE thick films contain micropores. An organic conducting polymer, poly(3,4-ethylenedioxythiophene): poly(styrenesulfonate) (PEDOT:PSS), was infiltrated into the micropores to achieve a hybrid composite material. Flexible TE modules were made, and the output voltage of the module is about 2 mV/K.

To independently tune the electrical conductivity and thermal conductivity, aerogels were studied to make TE energy conversion materials. Carbon nanotube (CNT) 3D networks connected by a resin were prepared followed by high-temperature calcination [5]. The heat treatment allows the formation of aerogels with nanoporous structures. The prepared lightweight aerogel materials with high porosity and high surface areas provide special thermal and electrical properties. It is found that the open-cell foam structure of such conducting aerogels suppresses the thermal conductivity significantly, while a good electrical conductive behavior is retained. Because the pyrolyzed aerogel is rigid, modification in the mechanical property of aerogels by using elastomers has been considered to make flexible TE generators. For example, PEDOT:PSS blended with a rubber poly(n-butylacrylate-styrene) (P(BA-St)) latex shows an elongation of up to 97% [6]. The p-type PEDOT:PSS has a figure of merit (*ZT*) value of 0.42 [7], which is a good candidate for making flexible TE devices running in the temperature range of 20°C to 250°C. PEDOT:PSS may be made into ink to be printed onto a flexible substrate [8].

Polymer TE modules have the advantage of being flexible. Aranguren et al. [9] designed a flexible TE energy generator based on the π-shaped PEDOT TE module. The TE module has been printed and tested. In comparison with the bismuth telluride commercial modules, the polymer shows lower energy conversion efficiency in a real application scenario of waste heat harvesting from a tile furnace with a smoke mass flow of 6.39 kg/s. Nevertheless, the polymer demonstrates the applicability to work at a temperature as high as 183°C. The alignment method of fibers and films was studied, and the anisotropies of PEDOT were obtained through the fiber alignment in a wet-spinning and hot-drawing process [10].

Two different TE effects, isobaric and nonisobaric, in conducting membranes were analyzed, and the relation between the isobaric Seebeck coefficient (*S*) and the nonisobaric Seebeck coefficient was derived [11]. It is found that the nonisobaric Seebeck coefficient has a larger variation than the isobaric Seebeck coefficient and can attain absolute values of above 10 mV/K. On the other hand, the isobaric Seebeck coefficient value is only around 1 mV/K. It is also predicted that large values of nonisobaric Seebeck coefficient may be found in charged/ion-conducting nanoporous membranes with pore sizes in the range of 5 to 200 nm [11].

Controlling the electrical conductivity of TE polymers is critical for improving the energy conversion efficiency. Menon et al. [12] studied the transport behavior of the n-type metallo-organic TE complex with nickel as the metal center and 1,1,2,2-ethenetetrathiolate (ETT) as the complex agent. It was found that the extent of oxidation can control the molecular composition and electrical conductivity of Ni-ETT to a large extent. The blend of Ni-ETT with PVDF attained an electrical conductivity value that is 20 times higher than the literature report at room temperature. Adding graphite to polymers is also found to be an effective way to increase the electrical conductivity (σ) as well as the Seebeck coefficient [13].

There are several kinds of conducting polymers most studied for TE module integration. Fang et al. [14] reported the rolled module design approach to integrate the p-type PEDOT:PSS and the n-type CNT composite. Nickel was also used as the n-type leg. To demonstrate the proof of concept, a rolled module comprising 288 legs was shown to produce an open circuit voltage of 260 mV and a power output of 46 μW at a temperature difference of 65 K. Conducting graphene/PANI nanocomposite fibers [15] that are flexible may also be used for the n-type leg. Menon et al. [16] introduced the radial TE module consisting of p-type and n-type conducting polymers. PEDOT/PSS with tellurium nanowires (NWs) was used as the p-type leg. Poly(nickel-1,1,2,2-ethenetetrathiolate) blended with poly(vinylidene fluoride)/dimethyl sulfoxide was used for making the n-type leg. Surface treatment using a new reagent of choline chloride (ChCl) urea mixture has been proposed for improving the TE performance of PEDOT:PSS [17]. There are simultaneous increases in the electrical conductivity and Seebeck coefficient of the polymer.

PANI has been studied as one of the promising TE polymers due to the considerably strong thermoelectricity [18]. It has good electrical conductivity [19], good stability in air [20], facile fabrication [21], low cost, high redox reversibility [22], and controllability of properties [23]. PANI has been investigated for application as battery electrodes [24], capacitors [25], chemical sensors [26], light-emitting diodes [27], anticorrosion reagents [28], electrochromic devices [29], and enzyme immobilizers [30]. PANI can be synthesized by the electrochemical and chemical approaches. They are both low-cost methods and can generate bulk PANI, thin films, and/or nanofibers. In the chemical process, residue surfactants and oxidants used possibly degrade the electrical property of the synthesized PANI. Therefore, the electrochemical method is considered as a better approach for fabricating cleaner PANI thin films or nanofibers because no surfactants and oxidants are needed in the process.

Similar to other TE materials, PANI has limited large-scale applications because of the unsolved decoupling issue among electrical conductivity, thermal conductivity, and Seebeck coefficient. Recently, combining organic and inorganic materials to make nanocomposites has been proposed to solve these problems. Conducting polymers as organic component are considered promising for enhancing the TE effect. The synthesized organic/inorganic nanocomposites could inherit the properties from both sides to achieve a synergistic effect. Moreover, organic/inorganic TE nanocomposites have the advantages of being low cost, easy to fabricate, and nontoxic and having a potentially high energy conversion efficiency. It is reported that Bi_2Te_3/PANI [31] and Sb_2S_3/PANI [32] have been successfully synthesized using electrochemical reactions with improved electrical conductivity. The Seebeck coefficients of inorganic materials have been reported in open literature. For example, the Seebeck coefficient of $DyCo_{0.95}Ni_{0.05}O_3$ is as high as 460 μV/K at 350 K [33]. Double perovskites Sr_2RuYO_6 and Sr_2RuErO_6 have the Seebeck coefficient of –250 μV/K at 1200 K [34]. The Seebeck coefficient of skutterudite $Yb_{0.26}Co_4Sb_{12}$/0.3GaSb is about –220 μV/K at 850 K [35]. However, the Seebeck coefficient data of organic/inorganic composites have been much less reported so far. This is the major motivation for the chapter: to introduce the results of the PANI conducting polymer on a nanostructured oxide surface.

5.2 Experimental Methods

To form TiO_2 NTs on a Ti substrate, a two-electrode cell was set up. A Ti foil substrate was the working electrode. Pt wire was used as the counterelectrode. The experiments were carried out at the room temperature of 22°C. The Ti substrate used was a rectangular titanium (50 mm × 4 mm) foil with a purity of 99.9% and a thickness of 0.5 mm. An electrolyte containing ethylene glycol, water, and ammonium fluoride was used to form TiO_2 NTs. The solution consisted of 90 wt% of ethylene glycol and 5 wt% of NH_4F, and the rest was distilled (DI) water. Before the electrochemical process, the Ti foil specimens were cleaned by isopropanol and DI water two times. After cleaning, the Ti foil specimens were immersed in the electrolyte, which had a depth of 25 mm, and the distance between the work electrode and the counterelectrode was 20 mm. The two electrodes were connected to a direct current power supply, and the Ti was the anode. The Pt served as the cathode. Morphology analysis by a scanning electron microscope shows that fully grown and well-aligned TiO_2 NTs form between 40 V and 50 V. The higher the voltage, the better the quality of the NTs is.

Specimens containing PANI on TiO_2 NTs were prepared after the Ti was electrochemically processed. The procedure for making the TiO_2 NTs/PANI nanocomposites is as follows: A three-electrode cell was set up first with the TiO_2 NT specimen as the anode. The Pt was the counterelectrode. The Ag/AgCl electrode was used as the reference electrode. An electrolyte consisting of 1 M hydrochloric acid, 0.2 M aniline (>99%), and DI water was made. Aniline was polymerized on the TiO_2 NT–containing Ti sheets with the processing parameters as shown in Ref. [18].

5.3 Results and Discussion

To measure the Seebeck coefficient of the TiO_2 NT/PANI samples, the two ends of each sample were bonded to strips of Al foils by using an Ag-based conductive adhesive. This would provide high electrical conductivity at the composite/electrode interface. Then one end of the Al foil was heated up to a required temperature (ranging from 30°C to 200°C in our experiments), while the other end was kept at an ambient temperature of 22°C.

The Seebeck coefficient measurement of the TiO_2 NTs/PANI composites indicates that the relatively high Seebeck coefficient of the composites is to a certain extent attributed to the low-dimensional TiO_2 NTs. Low dimensions confine the electrons to a narrow quantum well less than 20 nm in size. This leads to lower electron mobility in contrast to bulk materials. In such low-dimensional materials, the density of states (DOS) of electrons near the bottom of the conduction band and near the top of the valence band will increase when the quantum well size decreases. Theoretically, the increased DOS results in an enhanced Seebeck coefficient, as predicted by Hicks and Dresselhaus [36]. Experimentally, our recently published paper indicates that the low-dimensional TiO_2 NTs with Te-Bi-Pb nanoparticles show an improved Seebeck coefficient compared to pure TiO_2 NTs [37]. The highest absolute Seebeck coefficient value is 123.75 μV/K for the TiO_2 NT/PANI nanocomposite at 30°C. The Seebeck coefficient of the TiO_2 NT/PANI nanocomposite is higher than that of the pure TiO_2 NT specimen. The reason for the higher absolute Seebeck coefficient value of TiO_2 NT/PANI nanocomposite is the better conductive behavior of PANI. Therefore, PANI indeed contributes to the enhanced thermoelectricity of the composites. Typically, the Seebeck coefficient should increase with the increased temperature gradient. But in this case, the Seebeck coefficient was not enhanced with the increase in the temperature difference between the hot end and the cold end. It might be due to the faster thermal diffusion, resulting in the actual temperature difference between the hot and cold ends of the specimen being reduced more quickly at a higher thermal gradient.

5.4 Conclusions

Polymeric TE materials have been extensively studied recently due to the advantages they offer over metallic or ceramic TE materials. Polymeric materials have much lower densities and less thermal conductivity than metals or ceramics. In addition, polymers and their composites can be made flexible. Due to the varying molecular structures, polymers show tunable transport properties. The following families of conducting polymers and their composites are under active investigation: PEDOT, PANI, PTh, PPy, PA, and PC. Besides the synthesis work, TE energy conversion applications of these polymers have been studied.

PANI has been studied as one of the promising TE polymers due to the considerably strong thermoelectricity. It has good electrical conductivity and good stability, is easy to fabricate and low cost, has high redox reversibility, and allows one to control the transport properties. PANI can be synthesized by the electrochemical and chemical approaches. They are both low-cost methods and can generate bulk PANI, thin films, and/or nanofibers. The electrochemical method is considered a better approach to fabricate cleaner PANI thin films or nanofibers because no surfactants and oxidants are needed in the process. From the case studies of PANI/titanium dioxide NT hybrid composites, it is found that the TE response of the composites is more intensive than that of the pure titanium oxide NTs. The relatively high Seebeck coefficient of the composites is to a certain extent attributed to the low-dimensional

TiO_2 NTs. Low dimensions confine the electrons to a narrow quantum well less than 20 nm in size. This leads to lower electron mobility in contrast to bulk materials. The highest absolute value of the Seebeck coefficient for the TiO_2 NT/PANI nanocomposite materials was found to be 124 μV/K.

References

1. Gao, C., Chen, G. (2016). Conducting polymer/carbon particle thermoelectric composites: emerging green energy materials, *Compos. Sci. Technol.*, **124**, pp. 52–70.
2. Jundale, D. M., Navale, S. T., Khuspe, G. D., Dalavi, D. S., Patil, P. S., Patil, V. B. (2013). Polyaniline–CuO hybrid nanocomposites: synthesis, structural, morphological, optical and electrical transport studies, *J. Mater. Sci.: Mater. Electron.*, **24**, pp. 3526–3535.
3. Plochmann, B., Lang, S., Ruger, R., Moos, R. (2014). Optimization of thermoelectric properties of metal-oxide-based polymer composites, *J. Appl. Polym. Sci.*, **131**, p. 40038.
4. We, J. H., Kim, S. J., Cho, B. J. (2014). Hybrid composite of screen-printed inorganic thermoelectric film and organic conducting polymer for flexible thermoelectric power generator, *Energy*, **73**, pp. 506–512.
5. Zhao, L., Zhao, J., Sun, X., Li, Q., Wu, J., Zhang, A. (2015). Enhanced thermoelectric properties of hybridized conducting aerogels based on carbon nanotubes and pyrolyzed resorcinol–formaldehyde resin, *Synth. Met.*, **205**, pp. 64–69.
6. Khan, Z. U., Edberg, J., Hamedi, M. M., Gabrielsson, R., Granberg, H., Wågberg, L., Engquist, I., Berggren, M., Crispin, X. (2016). Thermoelectric polymers and their elastic aerogels, *Adv. Mater.*, **28**, pp. 4556–4562.
7. Aswal, D. K., Basu, R., Singh, A. (2016). Key issues in development of thermoelectric power generators: high figure-of-merit materials and their highly conducting interfaces with metallic interconnects, *Energy Convers. Manage.*, **114**, pp. 50–67.
8. Orrill, M., Saniya LeBlanc, S. (2017). Printed thermoelectric materials and devices: fabrication techniques, advantages, and challenges, *J. Appl. Polym. Sci.*, **134**, p. 44256.
9. Aranguren, P., Roch, A., Stepien, L., Abt, M. von Lukowicz, M., Dani, I., Astrain, D. (2016). Optimized design for flexible polymer thermoelectric generators, *Appl. Therm. Eng.*, **102**, pp. 402–411.

10. Patel, S. N., Chabinyc, M. L. (2017). Anisotropies and the thermoelectric properties of semiconducting polymers, *J. Appl. Polym. Sci.*, **134**, p. 44403.

11. Sandbakk, K. D., Bentien, A., Kjelstrup, S. (2013). Thermoelectric effects in ion conducting membranes and perspectives for thermoelectric energy conversion, *J. Membr. Sci.*, **434**, pp. 10–17.

12. Menon, A. K., Uzunlar, E., Wolfe, R. M. W., Reynolds, J. R., Marder, S. R., Yee, S. K. (2017). Metallo-organic n-type thermoelectrics: emphasizing advances in nickel ethenetetrathiolates, *J. Appl. Polym. Sci.*, **134**, p. 44402.

13. Wang, L., Wang, D., Zhu, G., Li, J., Pan, F. (2011). Thermoelectric properties of conducting polyaniline/graphite composites, *Mater. Lett.*, **65**, pp. 1086–1088.

14. Fang, H., Popere, B. C., Thomas, E. M., Mai, C. K., Chang, W. B., Bazan, G. C., Chabinyc, M. L., Segalman, R. A. (2017). Large-scale integration of flexible materials into rolled and corrugated thermoelectric modules, *J. Appl. Polym. Sci.*, **134**, p. 44208.

15. Ansari, M. O., Khan, M. M., Ansari, S. A., Amal, I., Lee, J., Cho, M. H. (2014). pTSA doped conducting graphene/polyaniline nanocomposite fibers: thermoelectric behavior and electrode analysis, *Chem. Eng. J.*, **242**, pp. 155–161.

16. Menon, A. K., Meek, O., Eng, A. J., Yee, S. K. (2017). Radial thermoelectric generator fabricated from n- and p-type conducting polymers, *J. Appl. Polym. Sci.*, **134**, p. 44060.

17. Zhu, Z., Liu, C., Jiang, Q., Shi, H., Xu, J., Jiang, F., Xiong, J., Liu, E. (2015). Green DES mixture as a surface treatment recipe for improving the thermoelectric properties of PEDOT:PSS films, *Synth. Met.*, **209**, pp. 313–318.

18. Su, L., Gan, Y. X. (2012). Experimental study on synthesizing TiO_2 nanotube/polyaniline (PANI) nanocomposites and their thermoelectric and photosensitive property characterization, *Composites Part B: Eng.*, **43**, pp. 170–182.

19. Park, C. H., Jang, S. K., Kim, F. S. (2018). Conductivity enhancement of surface-polymerized polyaniline films via control of processing conditions, *Appl. Surf. Sci.*, **429**, pp. 121–127.

20. Okamoto, H., Kotaka, T. (1998). Structure and properties of polyaniline films prepared via electrochemical polymerization. I: effect of pH in electrochemical polymerization media on the primary structure

and acid dissociation constant of product polyaniline films, *Polymer*, **39**(18), pp. 4349–4358.

21. Bai, S. L., Tian, Y. L., Cui, M., Sun, J. H., Tian, Y., Luo, R. X., Chen, A. F., Li, D. Q. (2016). Polyaniline@SnO_2 heterojunction loading on flexible PET thin film for detection of NH_3 at room temperature, *Sens. Actuators B: Chem.*, **226**(4), pp. 540–547.
22. Mu, S., Kan, J., Lu, J., Zhuang, L. (1998). Interconversion of polarons and bipolarons of polyaniline during the electrochemical polymerization of aniline, *J. Electroanal. Chem.*, **446**(1–2), pp. 107–112.
23. Zhang, J., Kong, L., Wang, B., Luo, Y., Kang, L. (2009). In-situ electrochemical polymerization of multi-walled carbon nanotube/polyaniline composite films for electrochemical supercapacitors, *Synth. Met.*, **159**(3–4), pp. 260–266.
24. Qiu, W., Zhou, R., Yang, L., Liu, Q. (1998). Lithium-ion rechargeable battery with petroleum coke anode and polyaniline cathode, *Solid State Ionics*, **86–88**(2), pp. 903–906.
25. Zhang, X., Ji, L., Zhang, S., Yang, W. (2007). Synthesis of a novel polyaniline-intercalated layered manganese oxide nanocomposite as electrode material for electrochemical capacitor, *J. Power Sources*, **173**(2), pp. 1017–1023.
26. Joubert, M., Bouhadid, M., Begue, D., Iratcabal, P., Redon, N., Desbrieres, J., Reynaud, S. (2010). Conducting polyaniline composite: from syntheses in waterborne systems to chemical sensor devices, *Polymer*, **51**(8), pp. 1716–1722.
27. Jang, J., Ha, J., Kim, K. (2008). Organic light-emitting diode with polyaniline-poly(styrene sulfonate) as a hole injection layer, *Thin Solid Films*, **516**(10), pp. 3152–3156.
28. Yang, X., Li, B., Wang, H., Hou, B. (2010). Anticorrosion performance of polyaniline nanostructures on mild steel, *Prog. Org. Coat.*, **69**(3), pp. 267–271.
29. Zhao, L., Zhao, L., Xu, Y., Qiu, T., Zhi, L., Shi, G. (2009). Polyaniline electrochromic devices with transparent graphene electrodes, *Electrochim. Acta*, **55**(2), pp. 491–497.
30. Crespilho, F. N., Lost, R. M., Travain, S. A., Oliverira Jr. O. N., Zucolotto, V. (2009). Enzyme immobilization on Ag nanoparticles/polyaniline nanocomposites, *Biosens. Bioelectron.*, **24**(10), pp. 3073–3077.
31. Chatterjee, K., Suresh, A., Ganguly, S., Kargupta, K., Banerjee, D. (2009). Synthesis and characterization of an electro-deposited polyaniline-

bismuth telluride nanocomposite: a novel thermoelectric material, *Mater. Charact.*, **60**(12), pp. 1597–1601.

32. Subramanian, S., Chithra Lekha, P., Pddiyan, D. P. (2010). Enhanced electrical response in Sb_2S_3 thin films by the inclusion of polyaniline during electrodeposition, *Physica B*, **405**(3), pp. 925–931.
33. Robert, R., Aguirre, M. H., Hug, P., Reller, A., Weidenkaff, A. (2007). High-temperature thermoelectric properties of Ln(Co, Ni)O_3 (Ln = La, Pr, Nd, Sm, Gd and Dy) compounds, *Acta Mater.*, **55**(15), pp. 4965–4972.
34. Aguirre, M. H., Logvinovich, D., Bocher, L., Robert, R., Ebbinghaus, S. G., Weidenkaff, A. (2009). High-temperature thermoelectric properties of Sr_2RuYO_6 and Sr_2RuErO_6 double perovskites influenced by structure and microstructure, *Acta Mater.*, **57**(1), pp. 108–115.
35. Xiong, Z., Chen, X., Huang, X., Bai, S., Chen, L. (2010). High thermoelectric performance of $Yb_{0.26}Co_4Sb_{12/y}GaSb$ nanocomposites originating from scattering electrons of low energy, *Acta Mater.*, **58**(11), pp. 3995–4002.
36. Hick, L. D., Dresselhaus, M. S. (1993). Effect of quantum-well structures on the thermoelectric figure of merit, *Phys. Rev. B*, **47**, p. 12727.
37. Su, L., Gan, Y. X. (2011). Formation and thermoelectric property of TiO_2 nanotubes covered by Te-Bi-Pb nanoparticles, *Electrochim. Acta*, **56**(16), pp. 5794–5803.

Chapter 6

Thermoelectric Properties of Bismuth Telluride–Filled Silicone

The objective of this chapter is to introduce how to manufacture flexible thermoelectric materials. A bismuth telluride–filled silicone rubber composite was extruded into millimeter-sized wires using electrospinning. The composite wires were tested in view of the electrical resistance and the Seebeck coefficient. The highest electrical resistance measured is 2.9×10^{10} ohms. The composite material exhibited a high Seebeck effect because silicone rubber exhibits low thermal conductivity as a result of increased phonon scattering. Moreover, compared with the bulk reference material, the thermoelectric property of bismuth telluride is notably enhanced. Due to the inherent flexibility of silicone rubber and the thermoelectric property of bismuth telluride, it is possible to make a flexible thermoelectric material for alternative energy applications.

6.1 Introduction

Thermoelectric (TE) materials have the capability to convert waste heat energy into useful electric energy. In recent years, much research has been done to observe this phenomenon. There are limitations associated with TE composite materials. It is a compromise between thermal conductivity and electrical conductivity of the TE material.

Nanomaterials for Thermoelectric Devices
Yong X. Gan

ISBN 978-981-4774-98-7 (Hardcover), 978-0-429-48872-6 (eBook)
www.panstanford.com

To optimize the Seebeck effect, thermal conductivity should be reduced whereas electrical conductivity is supposed to be enhanced. Furthermore, high-performance TE materials can be applied to recover waste heat from industry and vehicle tail pipes. The internal combustion process of an engine wastes a huge portion of the energy as waste heat. TE materials can be implemented to recover waste heat, and it is possible to achieve higher efficiency without any engine modification.

As a compound of tellurium and bismuth, bismuth telluride exhibits excellent TE property when it is alloyed with other elements, such as selenium or antimony. Bismuth telluride has been used for the construction of TE modules due to its high TE figure of merit [1]. The bulk bismuth telluride alloy is one of the most widely used commercial TE materials. It has a maximum figure of merit (ZT) close to unity at room temperature [2]. The TE figure of merit in a dimensionless form, ZT, has increased from 0.5 to greater than 1 in today's standard [1]. Z is the power factor. T is the absolute temperature. ZT is a function of the Seebeck coefficient (S), electrical conductivity (σ), thermal conductivity (κ), and absolute temperature (T). The equation is given by $ZT = S^2\sigma T/\kappa$. As described in the previous paragraph, a good TE material should have a high σ and a low κ. An excellent TE material should have a low thermal conductivity value. In general, increased scattering of heat carriers at interfaces lowers the κ value. Changes in the charge carrier density near the Fermi level could increase σ and S [3]. The Fermi level is the top collection of electron energy levels at absolute zero temperature, and the probability of occupation by electrons is half.

Bi_2Te_3-based solid solutions usually have ZT values close to 1. They are manufactured by unidirectional crystal growth methods, such as the Bridgeman or zone melting technique, that would lead to the preferred crystalline orientation [4]. Nevertheless, the high degree of texture is associated with weak van der Waals bonding between $Te^{(1)}$ and $Te^{(1)}$ layers of the quasi-layered crystalline structure, which results in poor mechanical properties [4].

Lattice thermal conductivity is inherently related to phonon transport [5]. In semiconductors and insulators, heat is transferred and quantified by vibrations in the crystal lattice. Inside a material, high frequencies are associated with heat while low frequencies correlate to sound. Part of the thermal conductivity κ comes from phonons traveling through the crystal lattice. Carrier contribution is related to the electrical conductivity σ according to the Wiedemann–Franz (WF) law [6]. The law says that raising the temperature would result in an increase in the thermal conductivity while decreasing the electrical conductivity. Often, electrical conductivity is easy to measure. However, the thermal conductivity is estimated by the WF law since the ratio of heat transport and charge transport by electrons is constant [7]. The total thermal conductivity is contributed by phonons and electrons. As a consequence, the thermal conductivity caused by phonons and that caused by electrons are tested separately [7]. The electron thermal conductivity is calculated by $\lambda_e = L\sigma T$, where the constant of proportionality is the Lorenz number. By definition, the total thermal conductivity comprises phonon thermal conductivity and electron thermal conductivity. Thus, the equation is given by $\lambda_{ph} = \lambda_{Total} - \lambda_e$ [7].

Bismuth telluride (Bi_2Te_3) is also a low-temperature TE material and is used in TE generators and coolers [8]. The electrical and thermal conductivities of bismuth telluride are anisotropic, which means it has different properties in different directions but the Seebeck coefficient does not depend on the orientation in the extrinsic or one-carrier regime [1]. The $Bi_2Te_{3-y}Se_y$ single-crystal bulk is a lamella structure, and the van der Waals bonding force between $Te^{(1)}$ and $Te^{(1)}$ causes easy cleavage along the planes perpendicular to the c axis. In contrast, n-type $Bi_2Te_{3-y}Se_y$ single-crystal solid solutions exhibit a strong anisotropic property [9].

The crystal structure of Bi_2Te_3 is rhombohedral, with the space group D^5_{3d} ($R\overline{3}m$). It has five atoms in the trigonal unit cell [10]. As a semiconductor, it has a 0.150 eV energy gap [11]. The energy gap refers to the difference between the valence band and the conduction band. As a narrow-band semiconductor compound, it is crystallized in the structure of tetradymite Bi_2Te_2S. Tetradymite also has a rhombohedral symmetry (space group $D^5_{3d} = R\overline{3}m$, structure type, C^3_3) [12]. The rhombohedral unit cell contains one Bi_2Te_3 molecule, but the structure is usually referred to as a hexagonal unit cell, consisting of 15 layers of 3 five-layer packets. The reason for referring to it as a hexagonal unit cell is that it is convenient to rearrange it into a hexagonal unit cell built of three formula units [13]. For instance, they are quintets $Te^{(1)}$-Bi-$Te^{(2)}$-Bi-$Te^{(1)}$-alternating along the hexagonal axis [12]. The closest neighbors of $Te^{(1)}$ are three

$Te^{(1)}$ atoms and three Bi atoms. In contrast, $Te^{(2)}$ has six Bi atoms that are the closest. The Te_2-Bi bond is inherently covalent, and the Te_1-Bi bond is mixed covalent and ionic [14]. The hexagonal cell contains 15 atoms grouped in 3 quintuple layers, as described from above. The rhombohedral unit-cell parameters are a_R = 10.473 Å and θ_R = 24.159 Å at 293 K. In addition, the hexagonal unit-cell parameters are a = 10.473 Å and c = 30.487 Å [15]. $Te^{(1)}$ and $Te^{(2)}$ are two different crystal kinds of tellurium atoms. For bismuth telluride, the hexagonal lattice parameters are $a\frac{\text{hex}}{\text{Bi}-\text{Te}}$ = 4.384 Å and $c\frac{\text{hex}}{\text{Bi}-\text{Te}}$ = 30.487 Å. For Be-Te_2, the closest interatomic distances between the individual monolayers inside the quintuple blocks are 3.25 Å [12]. However, the bond length of Te_1-Bi is 3.07 Å, which is shorter than the bond of Bi-Te_2 [15].

Polymers such as silicone rubber provide flexibility to the composite, but due to their insulating behavior, they do hinder the Seebeck effect [16]. However, the silicone rubber matrix is filled with semiconductor bismuth telluride. The presence of a silicone rubber matrix can increase phonon scattering and hence reduce thermal conductivity. Since silicone rubber possesses high heat resistance and thermal stability, it can survive in a large-temperature-gradient environment for TE applications. Silicone rubber compounds are made conductive by adding a conductive filler, such as carbon, metal powders, carbon fibers, carbon nanotubes, and graphite [17]. Despite the conductivity of carbon, it can potentially slow down the curing of rubber when it is used in high concentrations. There is a drawback when adding fillers to silicone rubber. The strength and flexibility of the composites will drop if the fillers are added in a large quantity. Another process that will enhance the conductivity of silicone rubber composites is heat treatment [18].

6.2 Experimental Methods

Bismuth telluride, 0.2 g, was mixed with 1.0 g of silicone rubber. The total weight was added to 1.2 g, and 0.12 g of the mixture was then transported to the syringe, which amounted to 10% of the bismuth telluride powder. In addition, the powder used in this experiment had 99.98% purity. The syringe needle used was a reusable stainless-steel dispensing needle from McMaster-Carr with a part number of 6710A41. The inside diameter of the syringe needle was 2.15 mm, and the outside diameter was 2.769 mm. The length of the syringe needle was 101.6 mm. As shown in Fig. 6.1, a syringe pump was instructed to pump the mixture at a rate of 0.1 mL per min. To achieve electrospinning, a DC power supply was implemented to deliver up to 30 kV of voltage to the syringe tip. However, severe discharging of electricity was observed when the voltage was set at 18 kV. Processed wires exhibited whipping rotations when the voltage was set at 16 kV. Then, the ground lead from the DC power supply was connected to the metal plate as shown in Fig. 6.1. Furthermore, the rectangular metal plate functioned as a grounded collector or a counterelectrode. Due to the high-voltage electric field, the bismuth telluride and silicone rubber mixture was stretched and extruded under electrostatic attraction and surface tension.

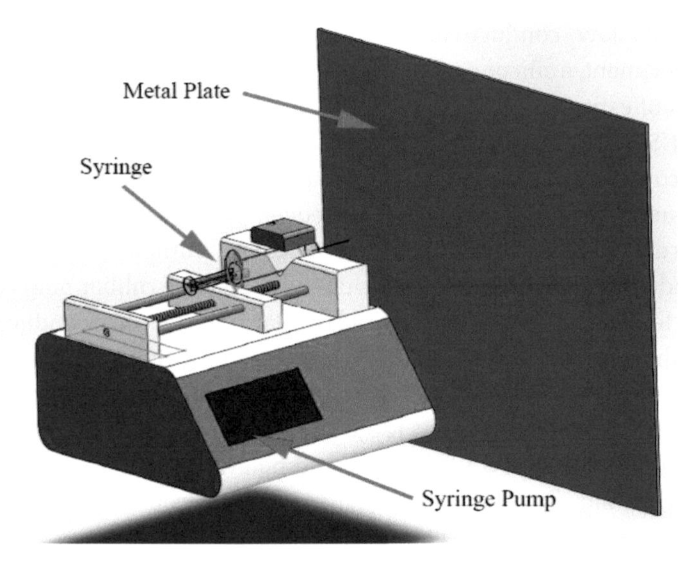

Figure 6.1 Schematic of a syringe connected to a power supply delivering DC voltage.

For the Seebeck coefficient measurements, a Talboys electrical heating plate was used to produce a temperature difference within the processed material. The rapid changes in temperature were measured by an infrared (IR) thermometer. To accommodate experimental accuracy and temperature gradient, one end of the sample was put in contact with the Talboys hot plate and the other end of the sample was left at the ambient temperature close to 25°C. The two ends of the bismuth telluride/silicone composite rod/wire specimens were wrapped with aluminum foils to ensure optimal electrical conductivity. The aluminum foil at each end served the purpose of creating an electrical conducting path for the electrical resistance and the Seebeck coefficient measurement. A CHI 600E electrochemical analyzer was used to record the current and the electrode potential. The resistivity and the Seebeck coefficient were calculated using the recorded data from the measurements.

6.3 Results and Discussion

Polymers such as silicone rubber possess low thermal conductivity and high electrical resistance. Pure polymeric materials usually exhibit low conductivity, less than 0.2 W/(m·K) [19]. In this experiment, a silicone matrix was used as an insulator. Two separate sets of experiments were performed to test for electrical resistivity and Seebeck coefficient. It was expected that the presence of the silicone rubber matrix would increase phonon scattering and hence reduce thermal conductivity. Moreover, the figure of merit would increase due to low thermal conductivity. Although the thermal conductivity of a bismuth telluride–filled silicone rubber matrix was not tested, the WF law can be used to relate the electrical conductivity to the electrical component of the thermal conductivity $\chi^{e}/\sigma T$ = constant = Lorenz number, where χ^{e} is electron thermal and σ stands for electrical conductivity [20]. The WF law equation is obtained by combining Eqs. 6.1 and 6.2, listed below. By combining electron charge and thermal conductivity derived from the Boltzmann transport equation, the equation becomes

$$\frac{k_{\mathrm{e}}}{\sigma}=\frac{\frac{1}{3}{v_{\mathrm{F}}}^{2}\tau_{\mathrm{F}}C_{\mathrm{e}}}{\frac{e^{2}}{m}n\tau_{\mathrm{F}}}=\frac{mC_{\mathrm{e}}{v_{\mathrm{F}}}^{2}}{3ne^{2}}. \tag{6.1}$$

Here, C_{e} is the electron specific heat, v_{F} is the Fermi velocity, τ_{F} is the scattering mean free time of Fermi electrons, n is the number density, m stands for mass, σ is the electrical conductivity, k_{e} is the thermal conductivity, and e is the electron charge [21, 22]. The electron specific heat C_e is derived from the thermal conductivity transport equation as

$$C_e = \frac{\frac{1}{2}\pi^2 n k_B T}{T_F} = \frac{\frac{1}{2}\pi^2 n k_B T}{\frac{m v_F^2}{2k_B}} = \frac{\pi^2 n k_B^2 T}{m v_F^2}, \tag{6.2}$$

where k_B is the Boltzmann constant, equal to 1.3807×10^{-23} m^2·kg·s^{-2}·K^{-1}, and T stands for temperature. Likewise, T_F is the Fermi temperature.

Thus, the ratio between thermal and electrical conductivity is described by

$$\frac{k_e}{\sigma} = \frac{m v_F^2}{3ne^2}\frac{\pi^2 n k_B^2 T}{m v_F^2} = \frac{\pi^2 k_B^2 T}{3e^2}. \tag{6.3}$$

The proportionality constant, Lorenz number, is defined by

$$L = \frac{\pi^2 k_B^2}{3e^2} = 2.45 \times 10^{-8} \frac{W\Omega}{K^2}. \tag{6.4}$$

Lastly, the WF law is defined by

$$\frac{k_e}{\sigma} = LT. \tag{6.5}$$

Two composites of bismuth telluride filled with silicone wires were tested for resistivity and Seebeck coefficient. The first sample was tested for resistance and I-V relation in two experimental trials as shown in Fig. 6.2 and Fig. 6.3. The voltammetry method used to test for resistance was linear sweep voltammetry, which indicates voltage was swept linearly with time between the working electrodes. The final voltage was set at 10 V, and the scan rate was adjusted at 0.1 V per second (Fig. 6.2 and Fig. 6.3). The same sample was tested two additional times for resistance to confirm accurate results. The test result showed no significant discrepancies in two consecutive experiment trials. In addition, analysis of several experimental trials shows that silicone rubber inherently exhibits high electrical resistance and the experiment showed that the silicone rubber control sample had an average resistance of 1.785×10^{10} ohms. The highest resistance tested in the first sample was 2.9×10^{10} ohms since bismuth telluride exhibited a topological insulating behavior and possessed overlapping conduction and valence bands, with a single nondegenerate surface-state band exhibiting a Dirac cone structure [23]. Unlike what happens when conductive fillers such as carbon and metal powders are added, filling bismuth telluride in the silicone matrix will increase the overall electrical resistance of the composite material. Since this was an experiment conducted to determine the Seebeck coefficient and the electrical resistance of the composite material, errors and uncertainties were inevitable and multiple trials of the experiment were performed to testify accuracy. Errors could result from false readings of the electrochemical analyzer CHI 600. When testing for the Seebeck coefficient, the composite material was heated using a Talboys heating plate. Due to a time delay when measuring the temperature difference between the opposite ends of the composite material, errors could result in the Seebeck coefficient data.

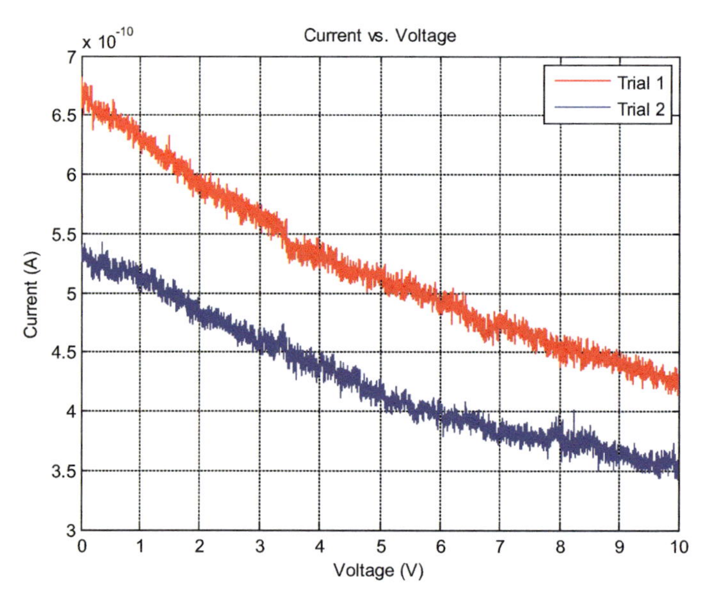

Figure 6.2 Current responses of the first composite under a supplied voltage with an increment of 0.001 V from 0 to 10 V.

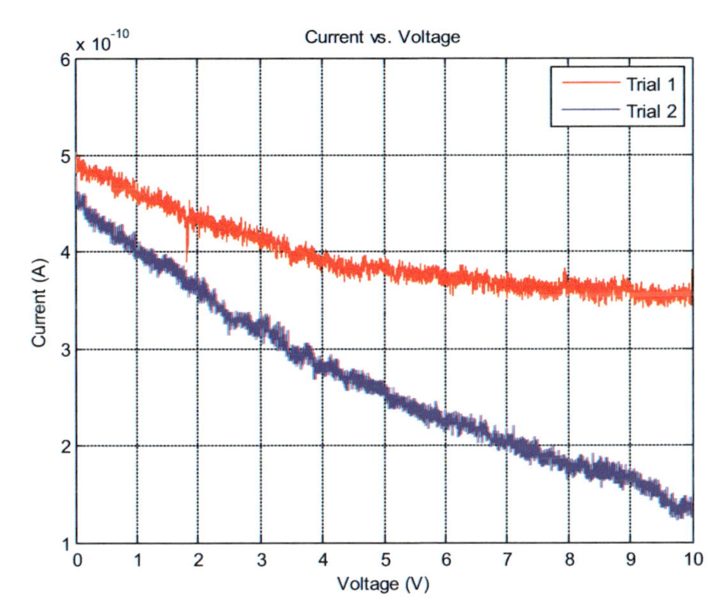

Figure 6.3 Current responses of the second composite under a supplied voltage with an increment of 0.001 V from 0 to 10 V.

To reduce thermal conductivity and increase the figure of merit, electrical resistance has to be created by using a silicone rubber matrix. Figure 6.4 presents the Seebeck coefficient of the material with varying hot-end temperatures at 313 K and 321 K. As the sample was heated on the Talboys electrical heating plate, the CHI electrochemical workstation took voltage data versus time. The temperature gradients were recorded simultaneously along with the voltage data. The output voltage in the sample was measured incrementally every 10 s within the time span of 150 s in the open circuit mode. The average Seebeck value at the hot-end temperature of 321 K was 0.01406 V/K, and the average Seebeck value was 0.029 V/K at 313 K. The measured values of bismuth telluride filled in a silicone matrix are significantly higher than the bulk reference value. It is apparent that the presence of bismuth telluride in the silicone rubber matrix can effectively enhance the TE property.

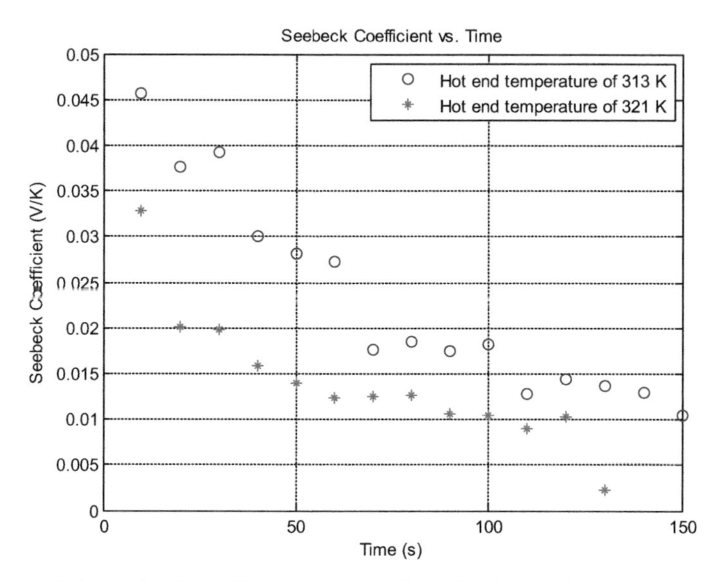

Figure 6.4 Seebeck coefficient measured at the hot-end temperatures of 313 K and 321 K.

6.4 Conclusions

In summary, the experiment investigated the TE property and electrical resistance of a bismuth telluride–filled silicone matrix through the electrospinning process. The highest electrical resistance measured in this experiment is 2.9×10^{10} ohms. The average measured resistance is 1.5325×10^{10} ohms. Although electrical resistance is increased by using a silicone rubber matrix, thermal conductivity is reduced as a result of the increase in phonon scattering. Hence, there is a compromise between electrical resistance and thermal conductivity when choosing silicone rubber as the matrix. As a result, the TE property of the composite material is improved significantly compared with the bulk reference material.

References

1. Goldsmid, H. J. (2014). Bismuth telluride and its alloys as materials for thermoelectric generation, *Materials*, **7**(4), pp. 2577–2592.

2. Mavrokefalos, A., Moore, A. L., Pettes, M. T., Shi, L., Wang, W., Li, X. G. (2009). Thermoelectric and structural characterizations of individual electrodeposited bismuth telluride nanowires, *J. Appl. Phys.*, **105**(10), Article No. 104318.
3. Purkayastha, A., Kim, S., Gandhi, D. D., Ganesan, P. G., Borca-Tasciuc, T., Ramanath, G. (2006). Molecularly protected bismuth telluride nanoparticles: microemulsion synthesis and thermoelectric transport properties, *Adv. Mater.*, **18**(22), pp. 2958–2963.
4. Xu, Z. J., Hu, L. P., Ying, P. J., Zhao, X. B., Zhu, T. J. (2015). Enhanced thermoelectric and mechanical properties of zone melted *p*-type $(Bi,Sb)_2Te_3$ thermoelectric materials by hot deformation, *Acta Mater.*, **84**(2) pp. 385–392.
5. Wang, Y. G., Qiu, B., McGaughey, A. J. H., Ruan, X. L., Xu, X. F. (2013). Mode-wise thermal conductivity of bismuth telluride, *J. Heat Transfer*, **135**(9), Article No. 091102.
6. Pinisetty, D., Devireddy, R. V. (2010). Thermal conductivity of semiconductor (bismuth-telluride)-semimetal (antimony) superlattice nanostructures, *Acta Mater.*, **58**(2), pp. 570–576.
7. Wang, H. D., Liu, J. H., Zhang, X., Takahasi, K. (2013). Breakdown of Wiedemann-Franz law in individual suspended polycrystalline gold nanofilms down to 3K, *Int. J. Heat Mass Transfer*, **66**(11), pp. 585–591.
8. Dheepa, J., Sathyamoorthy, R., Velumani, S., Subbarayan, A., Natarajan, K., Sebastian, P. J. (2004). Electrical resistivity of thermally evaporated bismuth telluride thin films, *Sol. Energy Mater. Sol. Cells*, **81**(3), pp. 305–312.
9. Yan, X., Poudel, B., Ma, Y., Liu, W. S., Joshi, G., Wang, H., Lan, Y. C., Wang, D. Z., Chen, G., Ren, Z. F. (2010). Experimental studies on anisotropic thermoelectric properties and structures of n-type $Bi_2Te_{2.7}Se_{0.3}$, *Nano Lett.*, **10**(9), pp. 3373–3378.
10. Mishra, S. K., Satpathy, S., Jepsen, O. (1997). Electronic structure and thermoelectric properties of bismuth telluride and bismuth selenide, *J. Phys.: Condens. Matter*, **9**(2), pp. 461–470.
11. Goncalves, L. M., Couto, C., Alpuim, P., Rolo, A. G., Völklein, F., Correia, J. H. (2010). Optimization of thermoelectric properties on Bi_2Te_3 thin films deposited by thermal co-evaporation, *Thin Solid Films*, **518**(10), pp. 2816–2821.
12. Pavlova, L. M., Shtern, Y. I., Mironov, R. E. (2011). Thermal expansion of bismuth telluride, *High Temp.*, **49**(3), pp. 369–379.

13. Yavorsky, B. Y., Hinsche, N. F., Mertig, I., Zahn, P. (2011). Electronic structure and transport anisotropy of Bi_2Te_3 and Sb_2Te_3, *Phys. Rev. B*, **84**(16), Article No. 165208.
14. Tong, Y., Yi, F. J., Liu, L. S., Zhai, P. C., Zhang, Q. J. (2010). Molecular dynamics study on thermo-mechanical properties of bismuth telluride bulk, *Comput. Mater. Sci.*, **48**(2), pp. 343–348.
15. Huang, B. L., Kaviany, M. (2008). *Ab initio* and molecular dynamics predictions for electron and phonon transport in bismuth telluride, *Phys. Rev. B*, **77**(12), Article No. 125209.
16. Kowalik, D., Chung, D. D. L. (2002). Carbon black filled silicone as a compliant thermoelectric material, *J. Reinf. Plast. Compos.*, **21**(17), pp. 1587–1590.
17. Joshi, A. M., Athawale, A. A. (2014). Electrically conductive silicone/organic polymer composites, *Silicon*, **6**(3), pp. 199–206.
18. Hu, S. F., Li, H., Chen, X. X., Zhang, C., Liu, Z. F. (2013). The electrical conductive effect of nickel-coated graphite/two-component silicone-rubber sealant, *Journal of Wuhan University of Technology-Mater. Sci. Ed.*, **28**(3), pp. 429–436.
19. Cheng, J. P., Liu, T., Zhang, J., Wang, B. B., Ying, J., Liu, F., Zhang, X. B. (2014). Influence of phase and morphology on thermal conductivity of alumina particle/silicone rubber composites, *Appl. Phys. A*, **117**(4), pp. 1985–1992.
20. Vekilov, Y. K., Isaev, E. I., Johansson, B. (2006). Does the Wiedemann–Franz law work for quasicrystals? *Phys. Lett. A*, **352**(6), pp. 524–525.
21. Westly, N. (2012). Electronic transport in thermoelectric bismuth telluride, University of New Orleans Theses and Dissertations, Paper 1539.
22. Arabshahi, H., Sarlak, F. (2010). A new study on calculation of electron transport characteristics in semiconductor materials, *Int. Arch. Appl. Sci. Technol.*, **1**(2), pp. 71–80.
23. Kioupakis, E., Tiago, M. L., Louie, S. G. (2010). Quasiparticle electronic structure of bismuth telluride in the GW approximation, *Phys. Rev. B*, **82**(24), Article No. 245203.

Chapter 7

Chemical Vapor Deposition of Complex Thermoelectric Materials

Complex materials have unique phonon and electron transport properties. In this chapter, a new manufacturing approach, catalyst-assisted metal organic chemical vapor deposition, was used to make Bi-Te-Ni-Cu-Au complex materials on an anodic aluminum oxide nanoporous substrate. Nickel acetate, copper nitrate, bismuth acetate, and tellurium (IV) chloride dissolved in *N,N*-dimethylformamide (DMF) were the metal sources for Ni, Bi, Cu, and Te, respectively. Hydrogen was used as the carrier gas. The anodic aluminum oxide substrate was sputter-coated with a gold thin film and was kept at 500°C in a quartz tube reaction chamber. The chemical vapor deposition time was 2 h. Scanning electron microscopic observation was performed to reveal the morphology of the deposited materials. The Bi-Te-Ni-Cu-Au materials showed self-assembled islands. The mechanism for morphology formation was proposed. The nanostructured coating has a higher Seebeck coefficient than most of the bulk materials.

7.1 Introduction

Chemical vapor deposition (CVD) is a bottom-up approach to make micro- or nanoscale materials. It is proposed to prepare low-

Nanomaterials for Thermoelectric Devices
Yong X. Gan

ISBN 978-981-4774-98-7 (Hardcover), 978-0-429-48872-6 (eBook)
www.panstanford.com

dimension thermoelectric (TE) materials [1]. The rate of CVD is controllable. For example, a high deposition rate was achieved to deposit n-type coating on a sapphire substrate [2]. Another feature of CVD is that the structures of deposited materials change with prepatterned metal dot catalysts [3, 4]. CVD has been used for depositing various TE materials, including silicon nanowires (Si NWs) [5, 6], SiC [7, 8], oxides [9], Ge nanocones [10], and Se-C films [11]. Metal-organic chemical vapor deposition (MOCVD) has caught a lot of attention for depositing Bi-Te and Sb-Te TE films [12–14]. It typically uses simple metal alkyl complexes as precursors.

The objective of this chapter (based on the recent work [15]) is to show how to make a Bi-Te-Ni-Cu-Au complex TE material on an anodic aluminum oxide (AAO) substrate by the MOCVD process. Nickel acetate, bismuth acetate, copper (II) nitrate, and tellurium (IV) chloride dissolved in *N,N*-dimethylformamide (DMF) were the metal sources for Ni, Bi, Cu, and Te, respectively. Hydrogen was used as the carrier gas. The structure of the deposited film was characterized by scanning electron microscopy. The mechanism of the microstructure development was discussed.

7.2 Materials and Experimental Method

The AAO template was purchased from Whatman, Inc., with a pore size of 0.2 μm and a diameter of 25 mm. The chemicals, nickel acetate, bismuth acetate, iron (III) nitrate, tellurium (IV) chloride, and DMF, were purchased from Alfa Aesar. The precursor solution was made by dissolving nickel acetate, bismuth acetate, copper (II) nitrate, and tellurium (IV) chloride in DMF in a 250 ml bottle. The nominal concentration for nickel acetate is 1.0 M. For copper (II) nitrate it is 0.5 M. The concentration of bismuth acetate is 0.1 M. For tellurium (IV) chloride, the concentration is 0.05 M. The solution was kept at 90°C. Hydrogen was inducted into the solution to carry the volatiles into a quartz reaction chamber (part 4 in Fig. 7.1) containing the AAO ceramic template (part 7 in Fig. 7.1) as the substrate for film deposition. The template was 0.5 mm in thickness and 12.5 mm in diameter. The AAO was sputter-coated with a 4 nm thick layer of Au using a model GSL-1100X-SPC12-LD mini plasma sputtering coater purchased from MTI Corporation. Figure 7.1 shows the configuration of the reactor. Before reaction, the quartz tube was vacuumed at a pressure level of 10.2 Torr. The hydrogen check valve was opened to sustain a volume flow rate of 5 standard cubic centimeters per minute (sccm). The furnace was heated up at the rate of 5°C/min. until it reached 500°C, using an MTI GSL-1100X-S50 split furnace (part 5 in Fig. 7.1). During the reaction, the exhaust gas was inducted into a container (part 9 in Fig. 7.1) filled with icy water, which allowed the DMF vapor to condense. The by-products from CVD reactions, such as hydrogen chloride and nitrogen oxide, were trapped in another container (part 10 in Fig. 7.1) and neutralized by a sodium carbonate solution. The remaining hydrogen was burned in a torch (part 11 in Fig. 7.1). After 1 h, the specimen was cooled down naturally.

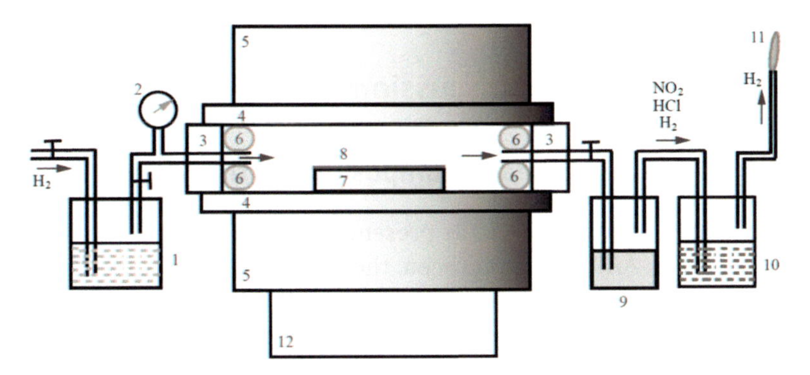

Figure 7.1 Schematic of the chemical vapor deposition setup. 1, precursor; 2, vacuum pump; 3, sealing flanges; 4, quartz tube; 5, split furnace; 6, thermal insulator; 7, AAO substrate; 8, Bi-Te-Ni-Cu-Au; 9, trap; 10, sodium carbonate solution; 11, gas torch; and 12, control unit [15].

The microstructure of the CVD coating was observed using a scanning electron microscope (JEOL JSM-6010PLUS/LA). Secondary electron images (SEIs) and backscattering electron composition (BES) images were taken. The SEI images were compared with the BES ones to examine the composition distribution. Quantitative analysis of elements was carried out using energy-dispersive X-ray spectroscopy (EDS).

Seebeck coefficient measurement was performed using a self-built facility. In brief, it consists of two functional units. One is the hot and cold temperature control unit. This unit was built on the basis of two major components, an aluminum heating ring and a digital temperature controller, both purchased from Columbia International Technical Equipment & Supplies LLC, Irmo, South Carolina. The other unit is for open circuit voltage measurement. The voltage measurement unit is made of a mode CHI440C Electrochemical Workstation and the accessories are purchased from CH Instrument, Inc., Austin, Texas. During the measurement process, one end each of both nanostructure specimens and the bulk material in the form of a thick coating was attached to the aluminum heating ring. The other end of the specimens was exposed to ambient temperature. The heating ring generated different temperature levels at the hot ends of the specimens. The value of Seebeck coefficient was calculated by the ratio of the voltage to the temperature difference between the hot ends and the cold ends of the specimens.

7.3 Results and Discussion

7.3.1 Morphology and Composition of the Material

The microstructure of AAO is presented. Figure 7.2a, a SEI, shows nanoporous AAO. As can be seen, the pore size is around 200 nm. The wall thickness between the pores is also around 200 nm. The energy-dispersive X-ray (EDX) diffraction spectrum reveals Al and O as the major elements. Scanning electron microscopic (SEM) analysis results of the microstructure and the composition of the deposited material are presented in Fig. 7.2b.

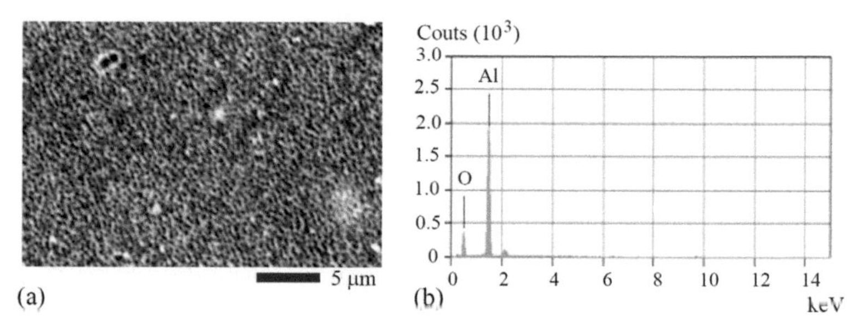

Figure 7.2 SEM image of AAO and composition profile: (a) SEI of the AAO and (b) EDS showing the qualitative results of Al and O elements [15].

A backscattered electron image for the deposited material is shown in Fig. 7.3a. In this image, the distribution of relatively heavy elements in the coating, such as Te and Bi, appears with a bright color. That of light elements is shown in black color. Figure 7.3a also shows several morphological features, including pores, islands, and clusters. These features are the evidence of a fast-growing mechanism of the complex TE material. Figure 7.3b shows the locations for spot analysis. Four representative locations were selected. The qualitative results obtained from the four sites are given in Figs. 7.3c, 7.3d, 7.3e, and 7.3f. The composition in mass percentage for each element from the four spots is presented in Table 7.1. The spectra in all four subfigures illustrate the diffraction peaks from five common elements: C, O, Al, Cu, and Te. A carbon signal comes from all four spots, revealing that carbon remains from the decomposition of DMF. On the one hand an oxygen signal can come from the template, AAO, and it is also a residual element from the decomposition of the DMF.

Table 7.1 Mass percentages of elements at the four spots [15]

Spot no.	001	002	003	004
O	16.05	12.61	17.64	12.21
Au	0	2.61	0	2.43
C	58.51	51.57	47.34	54.53
Mg	0	1.38	0	1.14
Al	23.66	18.91	32.97	16.76
Ni	0	1.34	0	1.53
Cu	0.6	1,1	0.58	1
Te	1.18	6.82	1.47	6.96
Bi	0	3.66	0	3.44

At Spot 001, the highest peak of Al is observed because it is the major element from the AAO template. Peak C is the second-highest one. Signal O follows them to generate the third-highest peak, indicating the dark shaded region is an AAO-rich region. At Spot 002, Al still shows the highest peak, while Au, Bi, and Ni are observed. Evidently, Au nanoparticles facilitated the formation of the complex Bi-Te-Al-Ni-Cu material. In addition, some Mg impurity signal is revealed at the spot site 002. At Spot 003, C, Al, and O appear as the major elements, indicating that the TE material growth was suppressed in this region as the case for Spot 001.

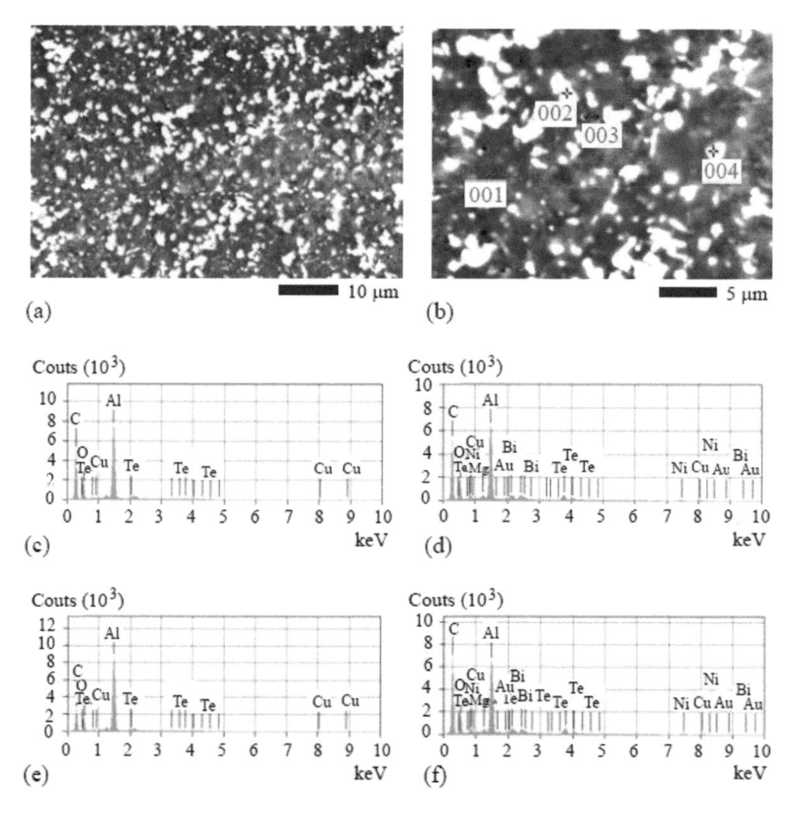

Figure 7.3 Spot analysis results: (a) BES image of the elements deposited on AAO, (b) BES image showing the locations for the EDS spot analysis, (c) EDS spectrum for Spot 001, (d) EDS of Spot 002, (e) EDS of Spot 003, and (f) EDS of Spot 004 [15].

To further validate the catalysis effect of the gold particle, Spot 004, with similar morphological features to those of Spot 002, was analyzed. Not surprisingly, Al still shows the highest peak. Au, Bi, and Ni are also found. This indicates that Au nanoparticles did result in the formation of the complex Bi-Te-Al-Ni-Cu material. Again, some minor amount of Mg impurity is found at Spot 004.

The quantitative results in mass percentages are listed in Table 7.1. For Spot 002 and Spot 004, the data in the last two columns yield the relative atomic ratio of Bi to Te, 3:10. This means that the deposited material is not a stoichiometric compound of Bi_2Te_3 as made by physical vapor deposition shown in Ref. [16]. The reason for such a difference is the low melting temperature of bismuth. Bismuth evaporates fast than telluride at a temperature above 250°C. This leads to the composition ratio less than 2:3 (the stoichiometric composition of the Bi-Te compound).

7.3.2 Nanostructure Growth Mechanism

The reaction mechanism of the gold particle–assisted MOCVD is studied. Earlier work has proven that gold thin films on polymer substrates possess the tendency to self-assemble into nanoscale dot or particle patterns at elevated temperatures [17]. In this work, a similar temperature effect should exist for the thin gold coating, leading to the formation of gold nanoparticles on the top surface of AAO. Gold nanoparticles have a high surface area and high activity. They became the active centers for the growth of the complex thermoelectric coating (TEC) material.

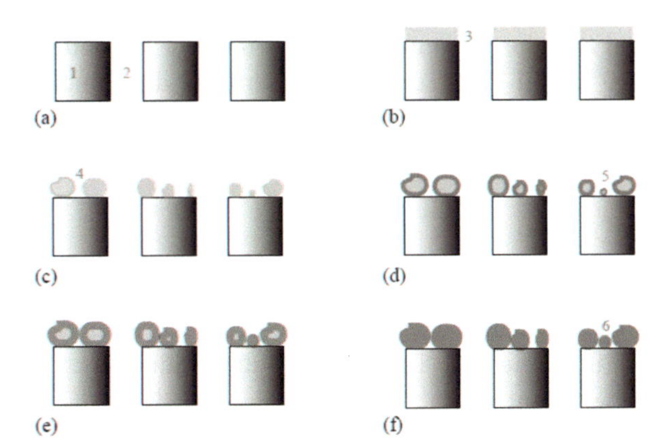

Figure 7.4 Schematic showing the gold catalyst–assisted chemical vapor deposition mechanism: (a) porous AAO, (b) AAO coated with a Au layer, (c) self-assembling of a Au layer into nanoparticles at elevated temperatures, (d) Bi-Te-Ni-Cu growth around Au, (e) reaction of Bi-Te-Ni-Cu with Au to form an alloy, and (f) homogenization of the complex material to form an island-like feature. 1, AAO wall; 2, AAO pore; 3, gold layer; 4, gold nanoparticles; 5, interface layer of Bi-Te-Ni-Cu and Au; 6, fully homogenized Bi-Te-Ni-Cu-Au coating [15].

The mechanism of gold nanoparticle–assisted catalytic growth is schematically shown in Fig. 7.4. Figure 7.4a shows porous AAO. Figure 7.4b shows the gold coating on it. When the coated AAO was set in the reaction chamber and heated up, due to the thermal mechanical effects, the coating shrank into particles to achieve the lowest free energy state possible, as shown in Fig. 7.4c. The gold nanoparticles became the centers for the TE material growth, as illustrated in Fig. 7.4d. With an increase in the processing time, more TE material got deposited at the interface, as presented in Fig. 7.4e. Homogenization of the material occurred, and the composition of the material became uniform through high-temperature diffusion, and the final product was just like that illustrated in Fig. 7.4f.

7.3.3 Thermoelectric Responses of Coating

First, the Seebeck coefficient results of a bulk Bi-Te-Ni-Cu-Au thick film material with a coating thickness of 200 nm are presented. The composition of the thick film is the same as that of the nanostructure, the only difference being that the deposition time was longer to form the thick and continuous film. The temperature differences between the hot end and the cold end were controlled to six values: 5°C, 6°C, 7°C, 8°C, 10°C, and 15°C. For each temperature difference condition, 4000 pieces of data were recorded, at the rate of 10 data pieces/s. Then the value of Seebeck coefficient, S, was calculated by using Eq. 7.1:

$$S = \frac{\Delta V}{\Delta T} \tag{7.1}$$

The mean value of each temperature difference condition was listed in Table 7.2. The final result obtained is the average Seebeck coefficient value S_{avg} = 105.85 ± 10.88 μV/K. This result is comparable with or slightly higher than the currently reported Seebeck coefficient value of the Bi-Te bulk material [18] or its composite material [19].

Figure 7.5 shows three typical curves for the Seebeck coefficient measurement results. These curves provide information of time-dependent TE responses of the thick coating. Figure 7.5a is a plot for the temperature difference of 15°C. Figure 7.5b demonstrates the trend for the specimen with the temperature difference of 10°C, and

Fig. 7.5c is the plot for the case of ΔT = 8°C. All the plots reveal the relatively stable TE response of the bulk material.

Table 7.2 Seebeck coefficient of a Bi-Te-Ni-Cu-Au thick film at various temperature differences between the hot end and the cold end [15]

Test no.	ΔT (°C)	S (μV/K)	Standard derivation (μV/K)
1	15	113.71	8.48
2	10	103.59	5.43
3	8	104.56	10.45
4	7	107.25	11.59
5	6	103.37	13.26
6	5	102.60	16.07
S_{avg} = 105.85 ± 10.88 μV/K			

The Seebeck effect tests on the nanostructure reveal meaningful results. As shown in Table 7.3, the results from seven tests consistently delivered the Seebeck coefficient value higher than 200 μV/K. The average value, S_{avg} = 237.14 ± 12.86 μV/K, is double the value of that for the thick film with the same composition, as shown by the results listed in Table 7.2.

Table 7.3 Seebeck coefficient of the Bi-Te-Ni-Cu-Au nanostructure [15]

Test no.	ΔT (°C)	S (μV/K)	Standard derivation (μV/K)
1	15	215.56	11.26
2	10	222.29	12.81
3	8	233.74	19.99
4	7	262.33	8.34
5	6	249.06	6.53
6	5	239.29	8.86
7	4	237.72	22.25
S_{avg} = 237.14 ± 12.86 μV/K			

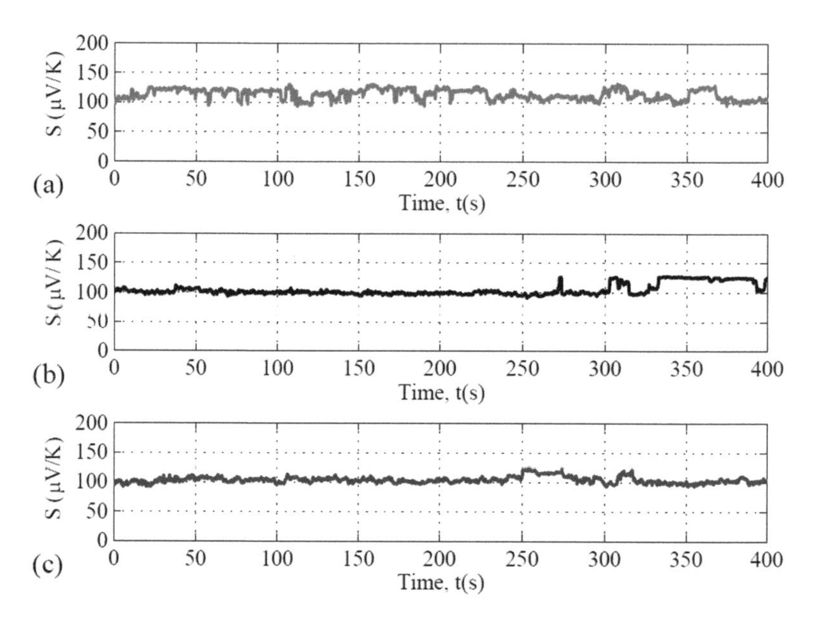

Figure 7.5 Seebeck coefficient measurement results of the thick coating obtained from various temperature differences between the hot end and the cold end: (a) with $\Delta T = 15°C$, (b) with $\Delta T = 10°C$, and (c) with $\Delta T = 8°C$ [15].

Figure 7.6 shows three typical curves for the Seebeck coefficient measurement results from the temperature difference of 5°C. Again, the three curves illustrate the time-dependent TE responses of the nanostructure. Figure 7.6a is a plot for the first test. Figure 7.6b reveals the TE response for the same specimen with the same temperature difference, 5°C, in the second test. Figure 7.6c is a plot for the third test, at ΔT = 5°C. These plots reveal the relatively stable TE response of the nanostructured material with the Seebeck coefficient values in the range of 230 to 260 µV/K.

The increase in the Seebeck coefficient of the nanostructure thin-film Bi-Te-Ni-Cu-Au is similar to the case for Bi-Te nanoparticles setting on silicon nanofibers as shown in our early work [20]. The enhanced Seebeck effect phenomenon can be explained by the quantum confinement in the Bi-Te nanostructure.

For such a low-dimensional nanostructure, the Seebeck coefficient is estimated by the Cutler–Mott formula [21]. Although this formula

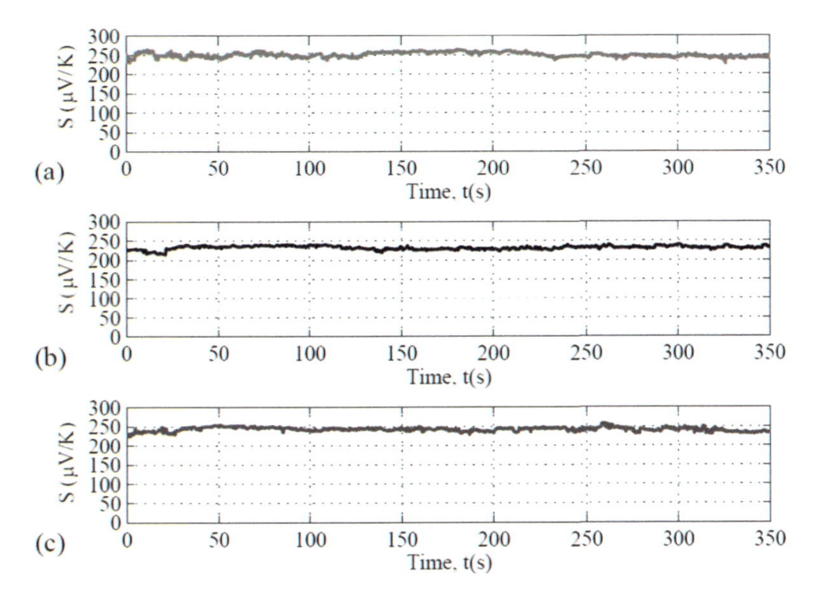

Figure 7.6 Time-dependent Seebeck coefficient results of the nanostructure obtained at ΔT = 5°C to show the repeated behavior of the nanostructure: (a) first test, (b) second test, and (c) third test [15].

was initially developed for degenerated semiconductors, it may also be applicable for the low-dimension structures as mentioned here for a preliminary evaluation. The Seebeck coefficient of the nanostructured Bi-Te-Ni- Cu-Cu alloy, S, can be expressed by

$$S = -\frac{\pi^2 k^2 T}{3e}\left[\frac{\partial ln\sigma(E)}{\partial E}\right]_{E=E_F}$$

$$\approx -\frac{\pi^2 k^2 T}{3e}\left\{\left[\frac{\partial ln\sigma(n)}{\partial E}\right]_{E=E_F} + \left[\frac{\partial ln\sigma(\tau)}{\partial E}\right]_{E=E_F}\right\}, \qquad (7.2)$$

where k is the Boltzmann constant, σ the conductivity, e the absolute value of the electron charge, n the density of charge carriers, τ the scattering time, and E_F the Fermi energy. The nanostructured particle/thin-film materials as prepared in this work were in the 1D and 2D form (islands and thin films in a rough AAO plane). The carrier concentration per unit area can be approximated by:

$$n = \int_0^{E_F} \frac{4\pi}{h^2} m_{\text{eff}} \left[\sum_{½} \Phi(E - E_v) \right] dE$$

$$= \frac{4\pi m_{\text{eff}}}{h^2} \left[\sum_{v} \Phi(E - E_v) \Phi(E - E_v) \right], \tag{7.3}$$

where v is the index of the discrete 1D or 2D energy levels, Φ the Heaviside step function whose value is 0 for a negative argument and equals 1 for a positive argument, m_{eff} the effective mass of the charge carriers, and E_v the vth energy level. E_v can be determined as:

$$E_v = -\frac{h^2 v^2}{8 d m_{\text{eff}}}, \tag{7.4}$$

where d is the diameter of the nanoparticles or the thickness of the thin film. From the above analysis, it is clear that the Seebeck coefficient should be increased by increasing the conductivity σ and the scattering time τ. The finer the nanostructure, or the thinner the film, the higher is the absolute value of E_v. Therefore, it is reasonable to see that the Bi-Te-Ni-Cu-Au nanostructure shows a stronger Seebeck effect than the bulk material in the thick film form. The Bi-Te-Ni-Cu-Au nanoscale islands are separated from each other on the surface of the AAO template. This is the mechanism for the enhancement of the Seebeck effect because of the reduced electron mobility from the islands. The implication of the research results in this work is that an inexpensive sensor may be built based on the highly sensitive TE responses of the Bi-Te-Ni-Cu-Au thin-film nanomaterial.

7.4 Conclusion

The Bi-Te-Ni-Cu-Au material is made by a one-pot precursor approach. Nickel acetate, bismuth acetate, copper (II) nitrate, and tellurium (IV) chloride in DMF are the metal sources for Ni, Bi, Cu, and Te, respectively. Hydrogen is used as the carrier gas to prevent oxidation of the TE material. The microstructure of the deposited material shows pores and island-like features. The islands are shown in the initial stage of the deposition, and such features are due to the existence of the noble metal (Au) nanoparticles. The potential applications of the deposited material, such as TE energy conversion and temperature sensing, are validated by the Seebeck coefficient measurement. The nanostructure shows a much higher value of the Seebeck coefficient than the thick film or the bulk material with the same composition. The reason for this is the quantum confinement of the nanostructure.

References

1. Huang, G., Jian, J., Lei, R., Cao, B. (2016). Single-crystalline Bi_2Se_3 nanowires grown by catalyst-free ambient pressure chemical vapor deposition, *Mater. Lett.*, **179**, pp. 198–201.
2. Tynell, T., Aizawa, T., Ohkubo, I., Nakamura, K., Mori, T. (2016). Deposition of thermoelectric strontium hexaboride thin films by a low pressure CVD method, *J. Cryst. Growth*, **449**, pp. 10–14.
3. Chen, P., Wang, J., Lu, Y., Zhang, S., Liu, X., Hou, W., Wang, Z., Wang, L. (2017). The fabrication of ReS_2 flowers at controlled locations by chemical vapor deposition, *Physica E*, **89**, pp. 115–118.
4. Haase, A., Peters, A., Rosiwal, S. (2016). Growth and thermoelectric properties of nitrogen-doped diamond/graphite, *Diamond Relat. Mater.*, **63**, pp. 222–226.
5. DaVila, D., Tarancon, A., Calaza, C., Salleras, M., Fernandez-Regulez, M., San Paulo, A., Fonseca, L. (2013). Improved thermal behavior of multiple linked arrays of silicon nanowires integrated into planar thermoelectric microgenerators, *J. Electron. Mater.*, **42**(7), pp. 1918–1925.
6. Wu, Y., Fan, R., Yang, P. (2002). Block-by-block growth of single-crystalline Si/SiGe superlattice nanowires, *Nano Lett.*, **2**(2), pp. 83–86.
7. Kim, J. G., Choi, Y. Y., Choi, D. J., Choi, S. M. (2011). Study on the thermoelectric properties of CVD SiC deposited with inert gases, *J. Electron. Mater.*, **40**(5), pp. 840–844.
8. Kim, J. G., Choi, Y. Y., Choi, D. J., Kim, J. I., Kim, B. S., Choi, S. M. (2009). A study on the thermoelectric properties of chemical vapor deposited SiC films with temperature and diluent gases variation, *J. Ceram. Soc. Jpn.*, **117**(1365), pp. 574–577.
9. Ngamou, P. H. T., Bahlawane, N. (2009). Chemical vapor deposition and electric characterization of perovskite oxides $LaMO_3$ (M=Co, Fe, Cr and Mn) thin films, *J. Solid State Chem.*, **182**(4), pp. 849–854.

10. Cho, H. S., Kamins, T. I. (2010). In situ control of Au-catalyzed chemical vapor deposited (CVD) Ge nanocone morphology by growth temperature variation, *J. Cryst. Growth*, **312**(16–17), pp. 2494–2497.
11. Grigorian, L., Fang, S., Sumanasekera, G., Rao, A. M., Schrader, L., Eklund, P. C. (1997). Physical properties of CVD-grown Se—carbon films, *Synth. Met.*, **87**(3), pp. 211–217.
12. Giani, A., Boulouz, A., Pascal-Delannoy, F., Foucaran, A., Boyer, A. (1998). MOCVD growth of Bi_2Te_3 layers using diethyltellurium as a precursor, *Thin Solid Films*, **315**(1–2), pp. 99–103.
13. Giani, A., Boulouz, A., Pascal-Delannoy, F., Foucaran, A., Boyer, A. (1999). Growth of Bi_2Te_3 and Sb_2Te_3 thin films by MOCVD, *Mater. Sci. Eng. B*, **64**(1), pp. 19–24.
14. Venkatasubramanian, R., Colpitts, T., Watko, E., Lamvik, M., El-Masry, N. (1997). MOCVD of Bi_2Te_3, Sb_2Te_3 and their superlattice structures for thin-film thermoelectric applications, *J. Cryst. Growth*, **170**(1–4), pp. 817–821.
15. Gan, Y. X., Yu, Z., Gan, J. B., Cheng, W., Li, M. (2018). Gold catalyst-assisted metal organic chemical vapor deposition of Bi-Te-Ni-Cu-Au complex thermoelectric materials on anodic aluminum oxide nanoporous template, *Coatings*, **2018**(5), pp. 8050166-1–8050166-12.
16. Zhang, G., Yu, Q., Wang, W., Li, X. (2010). Nanostructures for thermoelectric applications: synthesis, growth mechanism, and property studies, *Adv. Mater.*, **22**(17), pp. 1959–1962.
17. Do, M. T., Tong, Q. C., Lidiak, A., Luong, M. H., Ledoux-Rak, I., Lai, N. D. (2016). Nano-patterning of gold thin film by thermal annealing combined with laser interference techniques, *Appl. Phys. A*, **122**(4), p. 360.
18. Goldsmid, H. J. (2014). Bismuth telluride and its alloys as materials for thermoelectric generation, *Materials*, **7**(4), pp. 2577–2592.
19. Mahmoud, L., Alhwarai, M., Samad, Y. A., Mohammad, B., Laio, K., Elnaggar, I. (2015). Characterization of a graphene-based thermoelectric generator using a cost-effective fabrication process, *Energy Procedia*, **75**, pp. 615–620.
20. Gan, Y. X., Koludrovich, M. J., Zhang, L. (2013). Thermoelectric effect of silicon nanofibers capped with Bi–Te nanoparticles, *Mater. Lett.*, **111**, pp. 126–129.
21. Mott, N. F. (1987). *Conduction in Non-Crystalline Materials*, Oxford: Clarendon Press, p. 53.

Chapter 8

Porous Thermoelectric Materials Made by Infiltration Casting

This chapter deals with the thermoelectric effect of porous materials. Porous materials are supposed to have the property of slowing down heat flow. When a temperature gradient exists across a porous thermoelectric material, an electrical voltage is induced due to the difference in the energy levels of the electrons on the hot side versus the electrons on the cold side. An electrical voltage due to a temperature gradient exists due to the Seebeck effect. However, how to sustain the temperature gradient remains to be resolved. It is hypothesized that porous materials should have much lower heat conductivity than their bulk counterparts without sacrificing the electrical conductivity. In this study, three porous thermoelectric samples were produced by infiltration casting of bismuth tin (Bi-Sn) in a porous medium. The samples varied in their levels of porosity. The voltage output was measured for each sample at various temperature gradients over a period of 40 s at 0.1 s intervals. The Seebeck coefficient and the figure of merit (*ZT*) value of each sample were calculated. The average Seebeck coefficients for the three samples, from least porous to most porous, were -1.64×10^{-5}, -2.55×10^{-5}, and -1.54×10^{-5} V/K. The average figure of merit values for the three samples, from least porous to most porous, were 0.012, 0.031, and 0.010.

Nanomaterials for Thermoelectric Devices
Yong X. Gan

ISBN 978-981-4774-98-7 (Hardcover), 978-0-429-48872-6 (eBook)
www.panstanford.com

8.1 Introduction

According to the U.S. Department of Energy, about 67% of the energy consumed in the United States is lost as waste heat [1]. As a result, capture and conversion of waste heat to electricity has been a subject of much interest and research in recent years. Heat can be converted to electrical power by using thermoelectric (TE) devices. TE devices can also be used to generate cooling by electrical power, which is known as the Peltier effect. The advancement of TE materials in the last decade has fostered its use in many applications and industries. TE devices are used in telecommunication devices, home appliances, automobiles, and defense and space industries. They have also been the subject of much research on improvement in the efficiency of traditional power generating and refrigeration industries by capturing the waste heat and converting it to useful energy [2].

In the TE phenomenon a temperature difference across a conducting material yields a voltage output (electricity generation). This is known as the Seebeck effect [3]. The reverse is also possible, where a voltage potential can create a temperature difference. This is known as the Peltier effect [4]. When a conductor or a semiconductor is subjected to a temperature gradient, the electrons on the hot side have a higher thermal energy and will diffuse to the cold side [4]. As a result of the migration of charge carriers from the hot side to the cold side, they accumulate in the cold region, resulting in the formation of an electric potential difference [5]. This is known as the Seebeck effect and can be expressed by the following equation:

$$V = S(T_{\text{hot}} - T_{\text{cold}}), \tag{8.1}$$

where the voltage V is measured in volts, the Seebeck coefficient is measured in V/K, and the temperatures are measured in K. The equation above shows that the voltage produced is proportional to the Seebeck coefficient and the temperature gradient. The Seebeck coefficient will vary based on the type of TE material used [4]. When a voltage potential is applied to two electrodes of a TE material, a temperature gradient is created, which is known as the Peltier effect. Figure 8.1 shows the production of a temperature gradient when a voltage is applied. In this case, heat is absorbed on the left (cold junction) and is rejected on the right (hot junction) [4].

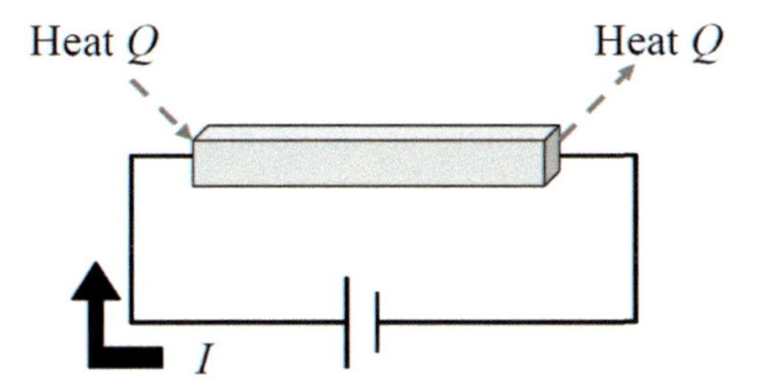

Figure 8.1 Demonstration of the Peltier effect, where a temperature difference is created when a voltage is applied to a thermoelectric material.

The Peltier effect is described by the following expression:

$$D = \Pi \times I, \tag{8.2}$$

where D is the rate of heat absorbed or rejected at the junctions, Π is the Peltier coefficient, and I is the current or electric charge flowing through.

The Seebeck and Peltier effects were both discovered in the 1800s [6]. Thomas Johann Seebeck is credited as one of the first discoverers of the basic TE effect, known as the Seebeck effect [7]. On December 14, 1820, at a session of the Berlin Academy of Sciences, Seebeck observed that a magnetic compass needle is deflected when the junction in a closed loop of two dissimilar metals are at different temperatures. Seebeck called this effect thermomagnetism [7]. Thirteen years after Seebeck discovered this effect, J. Peltier discovered the second of the TE effects. He discovered that if an electric current passes through a thermocouple a small heating or cooling effect is created [8].

In 1855, Lord Kelvin established the relationship between the Seebeck and Peltier effects by applying the theory of thermodynamics. He also showed that there is a third TE effect, which exists in homogeneous conductors. This effect involves reversible cooling or heating when there is a flow of current and a temperature gradient. This effect is known as the Thomson effect [8].

TE effects became a subject of much research in the 1950s, with the emergence of semiconductors and their alloys. However, they weren't widely used in the industry since they were not efficient enough to compete with the traditional mechanical compression cycle power generation and refrigerant equipment. In the 1990s, due to global energy crises and awareness of global warming and CO_2

emission, the quest for more environmentally friendly processes, such as TE devices, renewed the interest in thermoelectrics [6].

TE generators and coolers are solid-state devices without any moving parts. This is a major advantage since it increases reliability and decreases the chances of failure. The lack of moving parts also results in a device with no noise or vibration, which is a major advantage in many industries. TE devices are also small in size and very light in weight, which makes them more flexible in terms of their placement and integration with other devices and, therefore, very desirable for many industries and applications, such as aerospace [9]. For example, a silicon-germanium-based TE generator was used to power the spacecraft *Voyager* in the 1980s and is expected to be used for power generation in other spacecraft [10].

Since no ozone hole–creating chlorofluorocarbons (CFCs) or other refrigerants are used, TE devices are also considered to be environmentally friendly [9]. TE cooling is also used for cooling in electronics since they can provide localized cooling with a fast response. The major disadvantage of TE devices is their low efficiency. TE devices have lower efficiency when compared with electricity generation and cooling devices, which use traditional power, vapor, and refrigeration cycles [10]. However, the efficiency of the traditional devices can significantly increase when TE modules are strategically incorporated within their cycle.

In a recent study, a TE device was used as a dedicated subcooler at a condenser outlet to assist with improving the efficiency of the condenser. The use of the TE device resulted in a 16.2% increase in the system's coefficient of performance (COP) and increased the cooling capacity by 20% [2]. It must be noted that the increase in the capacity of a system with the addition of TE devices will also result in a reduced heat exchanger size, reducing the cost and size of the system [2].

8.2 Electric Current and Heat Flow Analysis

8.2.1 Heat Conduction

The current flow can be expressed by the following equation [11]. This equation assumes that the conductor is very long and so the contacts do not matter and the transport is diffusive [11].

$$J_{\mathrm{nx}} = \sigma_{\mathrm{n}} \frac{d(F_{\mathrm{n}}/q)}{dx}, \tag{8.3}$$

where σ_{n} is the electrical conductivity, which is equal to $1/\rho_{\mathrm{n}}$; F_{n} is the electromechanical potential (quasi-Fermi level); and $d(F_{\mathrm{n}}/q)/dx$ is the gradient of electromechanical potential.

We can also write the equation above in an alternate form, as follows [11]:

$$\frac{d(F_{\mathrm{n}}/q)}{dx} = \rho_{\mathrm{n}} J_{\mathrm{nx}}, \tag{8.4}$$

where ρ_{n} is the electrical resistivity.

When we have a temperature difference between two contacts the equation above can be written in the following form [11]:

$$J_{\mathrm{nx}} = \sigma_{\mathrm{n}} \frac{d(F_{\mathrm{n}}/q)}{dx} - S_{\mathrm{n}} \sigma_{\mathrm{n}} \frac{dT_{\mathrm{l}}}{dx} \tag{8.5}$$

or

$$\frac{d(F_{\mathrm{n}}/q)}{dx} = \rho_{\mathrm{n}} J_{\mathrm{nx}} + S_{\mathrm{n}} \frac{dT_{\mathrm{L}}}{dx}, \tag{8.6}$$

where S_{n} is the Seebeck coefficient in V/K and $dT_{\mathrm{L}}/\mathrm{dx}$ is the temperature gradient.

Equation 8.6 describes how much electromechanical potential gradient will result if we force an electrical current to pass through the material and impose a temperature gradient.

To relate the flow of current to the temperature gradient we must first develop an expression for heat flow in a material [11]. The heat flow can be expressed using Fourier's law as follows [12]:

$$J_{\mathrm{qx}} = -K \frac{dT}{dx}, \tag{8.7}$$

where J_{qx} is the heat flux (W/m^2), K is the thermal conductivity (W/m·K), and dT/dx refers to the temperature gradient (K/m).

Since the current flow through the material will also influence the heat flow in the material we must add another term to the heat flux equation to account for the current flow [11].

$$J_{\mathrm{qx}} = S_{\mathrm{n}} \sigma_{\mathrm{n}} T_{\mathrm{L}} \frac{d(F_{\mathrm{n}}/q)}{dx} - K \frac{dT}{dx} \tag{8.8}$$

The first term of Eq. 8.8 describes the heat flux due to the gradient in electromechanical potential, and the second term describes the heat flux due the temperature gradient.

Equation 8.8 can be simplified and written as follows [11]:

$$J_{qx} = \Pi_n J_{nx} - K_n \frac{dT_L}{dx}, \tag{8.9}$$

where $\pi_n = S_n T_L$, the Peltier coefficient W/A = V, and $K_n = K_{o-}(S_n{}^2 \sigma_n T_L)$ (W/m·K).

In Eq. 8.9, K_o is the thermal conductivity of the material when there is no current flow and the heat flow is strictly due to a temperature gradient. This term is commonly known as “short-circuit thermal conductivity.” K_n is the thermal conductivity of the material when there is heat flow due to a temperature gradient and a current flow. This term is known as “open-circuit thermal conductivity.”

8.2.2 Temperature Gradient and Voltage

When there is a temperature gradient between the two sides of a conductor, the electrons on the hot side have a higher thermal energy. As a result, the electrons will diffuse from the hot side to the cold side. The difference in charge concentration will result in a current flow and will induce a electrical voltage, which is known as the Seebeck effect [12], shown in Fig. 8.2. The voltage output V will be proportional to the temperature gradient. The proportionality constant is known as the Seebeck coefficient (V/K). The expression below shows the relationship described above [3].

$$V = S(T_{hot} - T_{cold}) \tag{8.10}$$

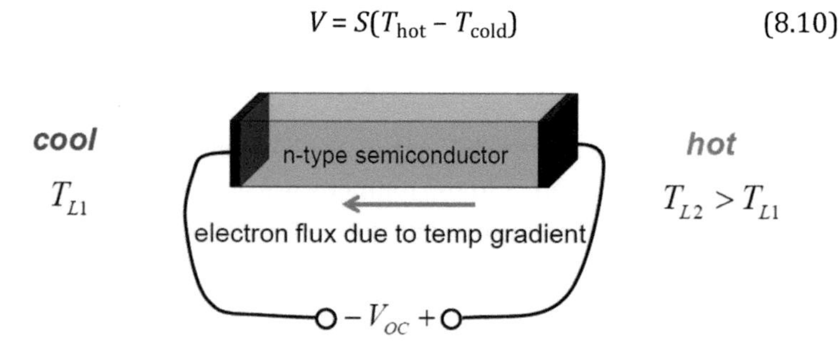

Figure 8.2 Development of voltage due to temperature gradient for an n-type semiconductor.

The direction of the voltage will vary based on the semiconductor type: n-type or p-type. The Seebeck coefficient for an n-type semiconductor will be negative and for a p-type semiconductor will be positive.

8.2.3 Efficiency of Thermoelectric Material

The electrical power output of a TE material is equal to V^2/R, where V is the voltage and R is the resistance of the material. This expression is proportional to σS^2, where σ is the electrical conductivity and S is the Seebeck coefficient. Consequently, since the heat flows from the hot portion to the cold portion due to a temperature gradient in the material, the heat flux is proportional to the thermal conductivity (κ). The efficiency of a TE device can be expressed as the ratio of the electrical power output to the heat flux input. Therefore, the efficiency of a TE material—for both cooling and electricity generation applications—can be defined by the following expression, which is known as the TE figure of merit [3]:

$$Z = \frac{\sigma S^2}{\kappa} \tag{8.11}$$

The expression above has the unit 1/K (inverse temperature). The figure of merit is commonly shown as a dimensionless number by multiplying both sides of the expression by the temperature [3].

$$ZT = \frac{\sigma S^2 T}{\kappa} \tag{8.12}$$

To achieve a high ZT value, the TE material would need to simultaneously have a high Seebeck coefficient and high electrical conductivity while also maintaining low thermal conductivity [6]. Since charge carriers—electrons in n-type and holes in p-type semiconductors—are responsible for carrying electric charge and heat, thermal conductivity and electrical conductivity have a proportional relationship [13]. The challenge in having a high figure of merit is the proportional relationship between electrical conductivity and thermal conductivity of metals [14].

If the temperature of a metal is held constant, thermal conductivity and electrical conductivity will have a proportional relationship.

This relationship is described by the Wiedemann–Franz (WF) law [13].

$$\frac{\kappa}{\sigma} = LT, \tag{8.13}$$

where κ is the thermal conductivity (W/m·K); σ is the electrical conductivity (1/Ωm); L stands for the Lorenz number, which equals 2.45×10^{-8} (W·Ω/K^2); and T is the temperature in Kelvin.

The Lorenz number in Eq. 8.13, which is the constant of proportionality, could vary on the basis of the type of material. However, most common materials will yield a similar number. Therefore, the Lorenz number is commonly treated as a constant number equal to 2.44×10^{-8} (W·Ω/K^2).

Another method of achieving a high figure of merit is to have a high Seebeck coefficient. This can be done by altering the band structure of the material by doping, which leads to an increase in the density of state to a level closer to the Fermi level [15].

As noted above, the efficiency of the TE material is expressed by the figure of merit. An ideal TE material will have a high Seebeck coefficient, high electrical conductivity, and low thermal conductivity. Since all of these properties are interdependent, optimizing one property will have a negative effect on the other. The challenge of producing an efficient TE material with a high figure of merit has been to increase the Seebeck coefficient and/or electrical conductivity while maintaining low thermal conductivity. Common TE materials, such as bismuth telluride (Bi_2Te_3) and lead telluride (PbTe), have ZT values close to 1. Using superlattices (SLs) and quantum dots on these materials has resulted in a reduction in the thermal conductivity without having a major effect on the TE power factor ($S^2\sigma$). The implementation of this technique has improved the ZT values of Bi_2Te_3 and PbTe to 2–2.4 [15].

8.2.4 Porosity and Thermal Properties of Porous Materials

In recent years, porous materials have found importance in many applications, such as filters, catalytic supports, and thermal insulators [16]. Porous materials are materials that contain voids (empty spaces) [3]. The amount of porosity in a material affects the structural, electrical, and thermal properties of the material. In this research, the effects of porosity on thermal properties of metals are of concern, specifically the effects on thermal conductivity. Heat transfer through a solid material is mostly a function of its thermal conductivity in a steady-state condition. The introduction of pores in a material significantly reduces its effective thermal conductivity and as a result improves the TE efficiency of a TE device [16].

The porosity of a material can be quantified by its true porosity and apparent porosity. Apparent porosity is the percentage of the total volume of the material that is occupied by its interconnected pores. True porosity is the percentage of the total volume of the material that is occupied by its interconnected and closed pores. Interconnected pores are pores through which gases and liquids can flow, and closed pores are pores that are closed and do not allow flow of liquids and gases [17].

Apparent porosity p_a can be calculated as follows:

$$p_a = \frac{W_w - W_d}{W_w - W_s} \times 100\%, \tag{8.14}$$

where W_d is the weight of the dry material, W_s stands for the weight of the material being suspended in water; W_w is the weight of the wet material after it has been taken out from water; and W_w - W_s measures the exterior volume of the sample, including the pores.

True porosity p_t can be calculated as follows:

$$p_t = \frac{\rho - B}{\rho} \times 100\%, \tag{8.15}$$

where B is the bulk density, which can be calculated by

$$B = \frac{W_d}{W_w - W_s} \times 100\%. \tag{8.16}$$

Bulk density is the ratio of the weight of the dry material to the exterior volume of the ceramic, including the pores. True density (ρ) is the density of material with no pores.

The closed pore percentage, m, can also be calculated by subtracting the apparent porosity from the true porosity.

$$m = p_t - p_a \tag{8.17}$$

The fraction of closed pores, *f*, can be calculated by dividing the closed pore percentage by the true porosity.

$$f = \frac{m}{p_t} \tag{8.18}$$

8.3 Experimental Procedures

8.3.1 Infiltration Casting Porous Bismuth Tin

The goal of this experiment was to produce a porous Bi-Sn material in order to analyze the TE properties of the Bi-Sn. The porous material was produced by infiltration casting. The experimental methodology is described below. The devices and setup for this experiment are shown in Fig. 8.3.

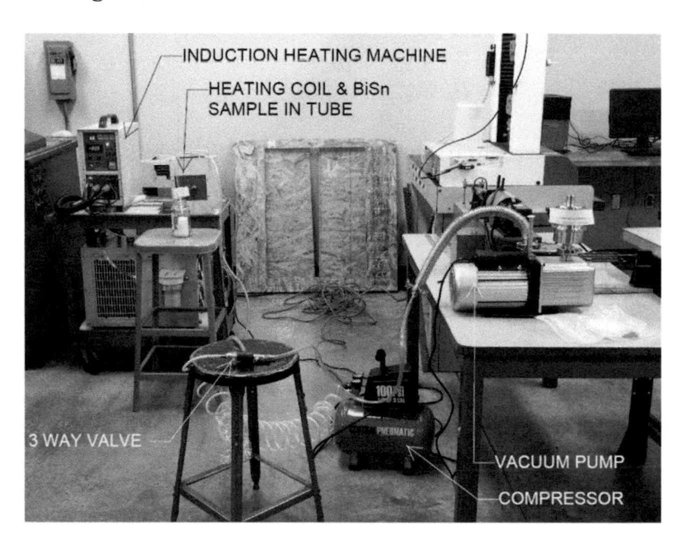

Figure 8.3 Photograph showing the infiltration casting experiment setup and equipment.

In Fig. 8.4, a quartz tube was first partially filled with pure salt (99.9% NaCl). Thinly cut pieces of Bi-Sn, which would be melted at a later stage, were placed on top of the salt. Due to high melting point of quartz, about 1700°C, a quartz tube was used to melt the material. The quartz tube used in this experiment had an outside diameter of 10 mm and an inside diameter of 8 mm. A plastic tube was used to connect the quartz tube to a vacuum pump and a compressor. A three-way manual valve was used to connect the compressor and the vacuum pump to the tube. An induction heating device (Magtech, split-type model EQ-SP-15AB) with a recirculation water chiller (MTI corporation model KJ-5000DI) was used to melt the metal. The induction heating coil was placed at the appropriate height so that it surrounded the Bi-Sn pieces for the highest heating efficiency.

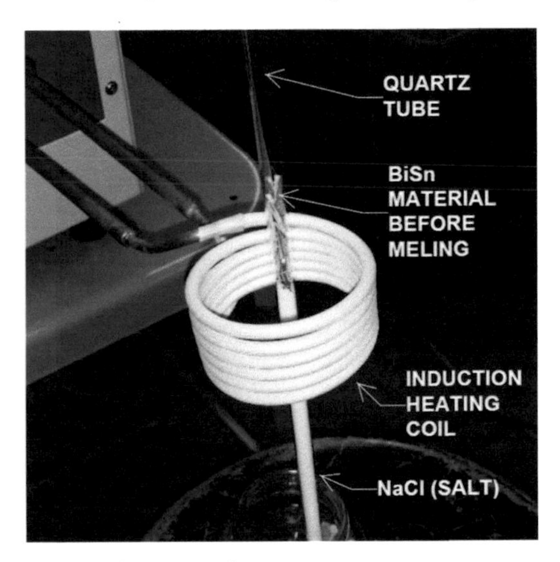

Figure 8.4 Induction heating coil setup.

First, the vacuum pump as shown in Fig. 8.5 was used to create a vacuum in the tube. The purpose of creating a vacuum was to keep the molten specimen afloat on top of the salt until the entire specimen was melted and ready to be shot by high-pressure air. At this point, the three-way valve was open on the vacuum and tube side and closed on the compressor side. Once vacuum had been created in the tube, the induction heating device was turned on to melt the metal.

Once the metallic alloy (Bi-Sn) had been completely melted, the compressor line, as shown in Fig. 8.6, which was held at 25 psi, was opened and the vacuum line was closed. The pressure forced the molten metal to infiltrate the salt, creating a metal filled with small salt grains.

Figure 8.5 9.5 CFM, two-stage vacuum pump (Model 2TW-4C).

Figure 8.6 Central pneumatic compressor, 1/3 hp, 3 gal.

After the metal was allowed to cool down, the material was removed from the tube and placed in water. The water dissolved the salt within the metal, and the resulting specimen was a porous material. The schematic representation of the experiment described above is shown in Fig. 8.7.

8.3.2 Voltage Output Measurement for Seebeck Coefficient Calculations

This subsection describes the procedure for measuring the voltage output given a specific temperature gradient at room temperature (25°C). Two sides of the sample (porous Bi-Sn) were connected to an acoustic wave microsensor. A CH instrument (model #

CHI440C), which is a time-resolved electrochemical quartz crystal microbalance (EQCM), was used. The microsensor, among other capabilities, is able to measuring potential or voltage output with respect to time. The microsensor is also equipped with high-speed data acquisition circuitry that allows for high-speed and high-frequency data recording, as shown in Fig. 8.8. In this experiment the voltage output was recorded over a 40 s period for each sample at increments of 0.1 s.

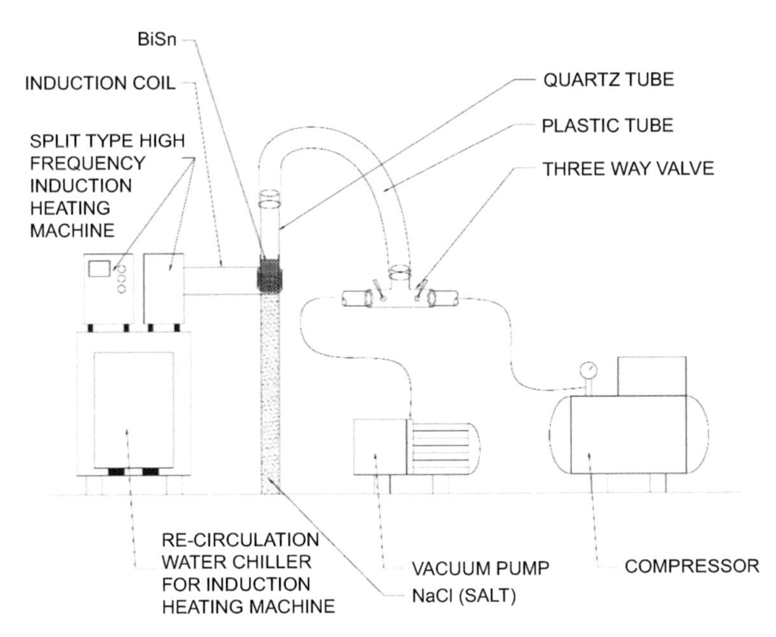

Figure 8.7 Schematic representation of an experiment for infiltration casting of a porous thermoelectric sample.

Aluminum foil was used to create a connection from the sample to the electrode leads of the microsensor, as shown in Fig. 8.9. Three electrode lead outputs were connected to our sample. Two electrodes were connected to one side of the sample and one to the opposite side. One of the electrodes was used as a reference for the voltage measurement. It must be noted that prior to the actual voltage measurement the device was calibrated by connecting the electrodes to an aluminum bar, which resulted in a zero-voltage output.

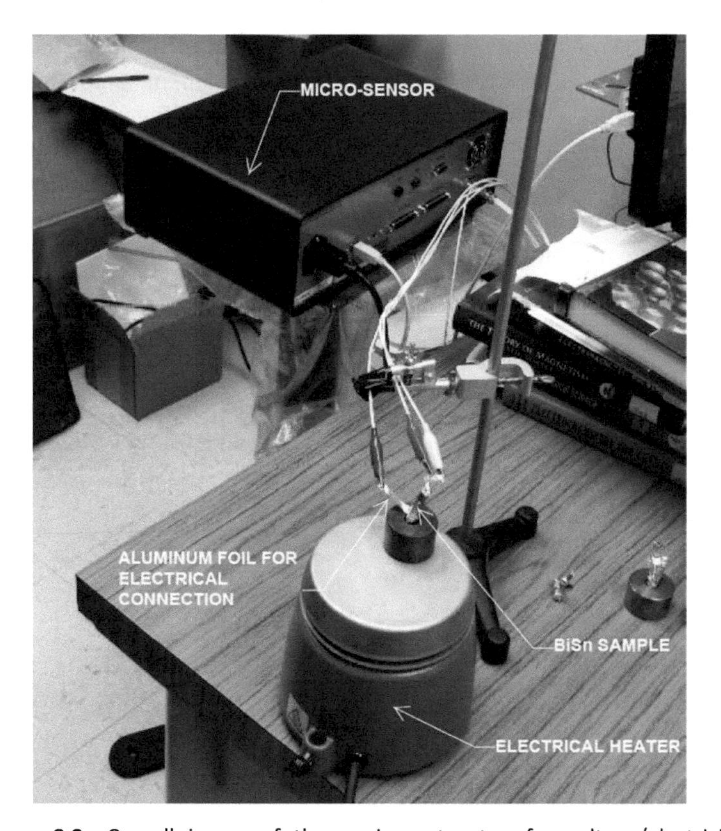

Figure 8.8 Overall image of the equipment setup for voltage/electricity generation.

Once the sample was connected to the microsensor, one side of the sample was placed on a circular piece of metal placed on an electric burner. The purpose of the piece of metal was to use its thermal mass to stabilize the temperature output of the electric heater because the surface temperature of the electrical heater fluctuated during the experiment.

The electrical heater was then energized, creating a temperature gradient on the TE Bi-Sn sample. The side of the sample that rested on the heated metal was the hot side, and the side exposed to the ambient temperature was the cold side. The temperatures on the hot and cold sides were then recorded using a laser thermometer.

The voltage output of the TE Bi-Sn sample was measured on the microsensor for three samples, at three various temperature gradients per sample. The voltage outputs for each test run were recorded for 39 s at increments of 0.1 s and plotted versus time. The resulting voltage outputs for each run were also used to calculate the Seebeck coefficient and figure of merit for each sample. The schematic representation of the procedure described above is shown in Fig. 8.10.

Figure 8.9 Connection of sample to microsensor for voltage measurement.

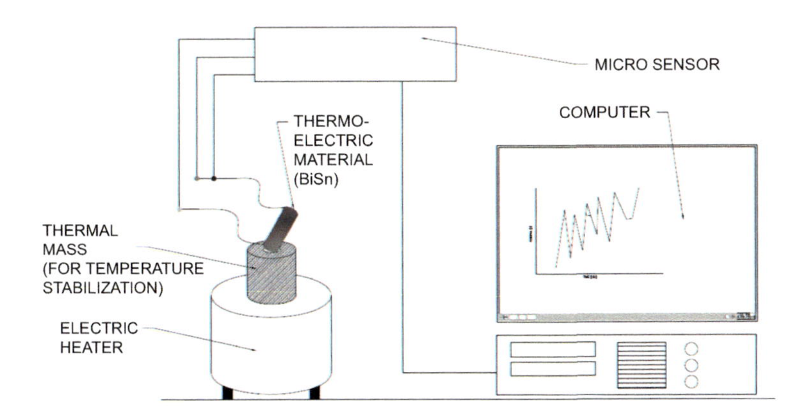

Figure 8.10 Schematic representation of the experiment for voltage/ electricity generation.

8.4 Results and Discussion

8.4.1 Morphology of Infiltration Cast Porous Bismuth Tin Alloy

In this experiment three porous samples of Bi-Sn were prepared, as described in Section 8.3.1. The samples are shown in Fig. 8.11 and were photographed with the aluminum contacts used to connect the samples to the electrodes of the microsensor for voltage output measurement.

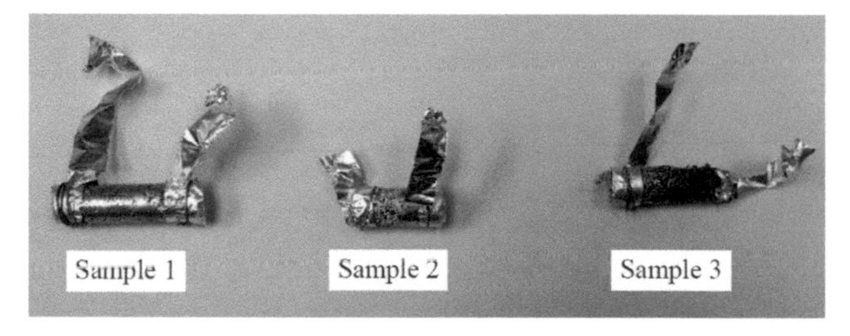

Figure 8.11 Samples of porous Bi-Sn produced in the experiment.

As illustrated by Fig. 8.11, sample 1 was the least porous of the samples. Samples 2 and 3 were medium porous and most porous, respectively. These samples were used as TE material to generate voltage by means of exposing the material to a temperature gradient. The Seebeck coefficient and figure of merit of these materials were also calculated and analyzed.

8.4.2 Voltage, Seebeck Coefficient, and Figure of Merit

8.4.2.1 Output voltage of each sample

The samples shown above were used to measure the voltage output given a specific temperature gradient using the procedure described in Section 8.3.2. The voltage output for each sample was measured—using a CH instruments microsensor (Model #CHI1410C) at three different temperature gradients. The microsensor recorded the voltage output at increments of 0.1 s. Various temperature gradients for each sample are shown in Tables 8.1, 8.2, and 8.3. The voltage outputs for each sample were also plotted versus time for 39 s at increments of 0.1 s. The plots of voltage output versus time are shown in Figs. 8.12–8.14.

Table 8.1 Seebeck and *ZT* calculations for sample 1 at three temperature gradients

Sample 1			
T hot (°C)	42	39	56
T cold (°C)	37	30	42
ΔT (K)	5	9	14
Voltage (V)	−1.04E-04	−1.51E-04	−1.62E-04
Seebeck coeff. (V/K)	−2.08E-05	−1.68E-05	-1.16E-05
Figure of merit (*ZT*)	0.018	0.012	0.005

Notes:

- For better accuracy, the voltage used in calculations are the average voltage output of the first ten seconds.
- The figure of merit was calculated from $ZT = \sigma S^2 T/K$. The values for thermal conductivity K, electrical conductivity σ, and temperature T are as follows:
  K (W/mK) = 9.639
  σ (1/Ωm) = 1.326E + 06
  T (K) = 298.00

Table 8.2 Seebeck and *ZT* calculations for sample 2 at three temperature gradients

Sample 2			
T hot (°C)	34	37	39
T cold (°C)	30	28	29
ΔT (K)	4	9	10
Voltage (V)	−1.61E-04	−1.68E-04	−1.76E-04
Seebeck coeff. (V/K)	−4.03E-05	−1.86E-05	−1.76E-05
Figure of merit (*ZT*)	0.067	0.014	0.013

Notes:

- For better accuracy, the voltage used in calculations are the average voltage output of the first ten seconds.
- The figure of merit was calculated from $ZT = \sigma S^2 T/K$. The values for thermal conductivity K, electrical conductivity σ, and temperature T are as follows:
  K (W/mK) = 9.639
  σ (1/Ωm) = 1.326E + 06
  T (K) = 298.00

Table 8.3 Seebeck and ZT calculations for sample 3 at three temperature gradients

Sample 3			
T hot (°C)	36	38	42
T cold (°C)	24	25	27
ΔT (K)	12	13	15
Voltage (V)	−1.96E-04	−1.85E-04	2.36E-04
Seebeck coeff. (V/K)	−1.63E-05	−1.43E-05	−1.57E-05
Figure of merit (ZT)	0.011	0.008	0.010

Notes:

- For better accuracy, the voltage used in calculations are the average voltage output of the first ten seconds.
- The figure of merit was calculated from $ZT = \sigma S^2 T/K$. The values for thermal conductivity K, electrical conductivity σ, and temperature T are as follows:
  K (W/mK) = 9.639
  σ (1/Ωm) = 1.326E + 06
  T (K) = 298.00

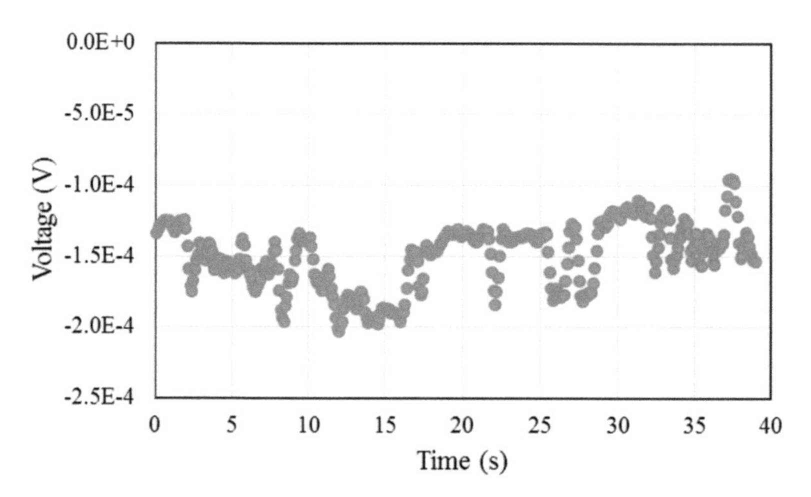

Figure 8.12 Plot of voltage versus time for sample 1 at ΔT = 9°C.

8.4.2.2 Seebeck coefficient calculations

The voltage outputs for each sample were then used to calculate the Seebeck coefficient for each material at that temperature gradient. The Seebeck coefficient was calculated using the following expression [3]:

$$S = \frac{V}{T_{\text{hot}} - T_{\text{cold}}} = \frac{V}{\Delta T} \tag{8.19}$$

The results of the Seebeck coefficient calculations for samples 1, 2, and 3 are tabulated in Tables 8.1, 8.2, and 8.3, respectively.

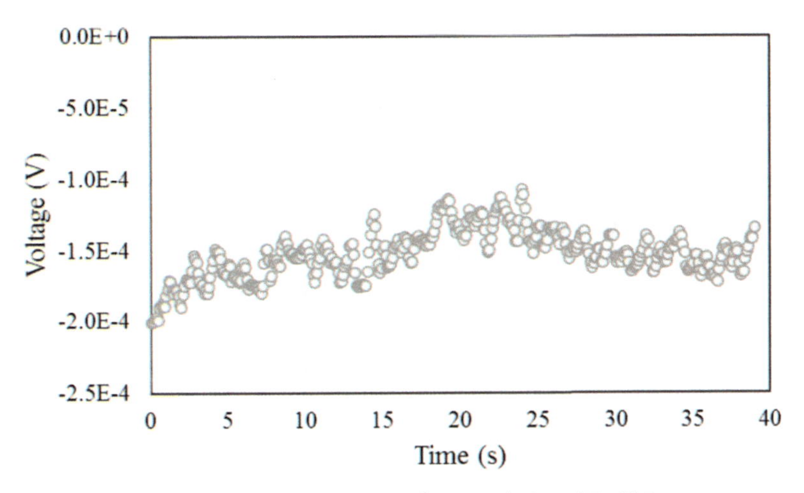

Figure 8.13 Plot of voltage versus time for sample 2 at ΔT = 9°C.

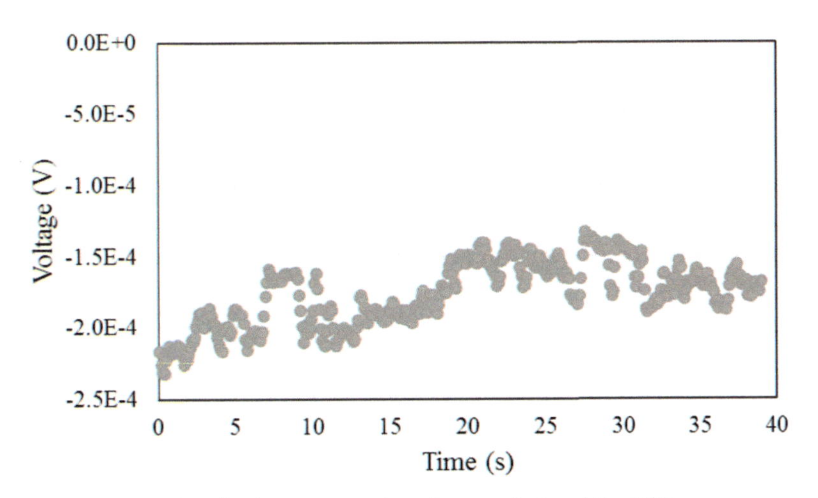

Figure 8.14 Plot of voltage versus time for sample 3 at ΔT = 12°C.

As discussed in Section 8.2, the Seebeck coefficient for an n-type semiconductor will be negative and for a p-type semiconductor positive. The results of the Seebeck coefficient calculations resulted in a negative number; hence, our samples were all n-type.

8.4.2.3 Figure of merit calculations

The figure of merit for each sample at a given temperature gradient was also calculated using the following expression [3]:

$$ZT = \frac{\sigma S^2 T}{\kappa} \tag{8.20}$$

The results of the ZT calculations for samples 1, 2, and 3 are tabulated in Tables 8.1, 8.2, and 8.3, respectively. The electrical conductivity value for our sample (59.20Bi-40.80) was estimated from a plot of electrical resistivity as a function of temperature obtained from literature [18]. Our experiment was performed at ambient temperature of 25°C. As can be seen from the plot in Ref. [18], the electrical resistivity has a linear relationship with respect to temperature. Using this linear relationship, the function of the line for 59.20Bi-40.80Sn was found. Using equation of the line, electrical resistivity at the desired temperature of 25°C was estimated to be $\rho = 75.433 \times 10^{-8}$ (Ωm). Since electrical resistivity has an inverse relationship with electrical conductivity, electrical conductivity was calculated as follows:

$$\sigma = \frac{1}{\rho} = \frac{1}{75.433 \times 10^{-8}} = 1.326 \times 10^{6}\ (\Omega^{-1} \cdot \mathrm{m}^{-1}) \tag{8.21}$$

As discussed in Section 8.3, thermal conductivity and electrical conductivity have a proportional relationship, which is described by the WF law [13]. This law relates electrical conductivity and thermal conductivity by the following expression:

$$\frac{\kappa}{\sigma} = LT, \tag{8.22}$$

where κ is the thermal conductivity, σ is the electrical conductivity, T is the temperature, and L is the Lorenz number (proportionality constant), which is equal to 2.44×10^{-8} (WΩ/K^2). Using the expression above the thermal conductivity was calculated as follows:

$$\kappa = LT\sigma = 2.44\times10^{8}\left(\frac{W\Omega}{K^{2}}\right)\times298(K)\times1.326\times10^{6}\left(\frac{1}{\Omega\cdot m}\right)$$

$$=9.639\left(\frac{W}{K\cdot m}\right) \tag{8.23}$$

8.4.2.4 Calculated values of Seebeck coefficient and figure of merit

The Seebeck coefficient and the figure of merit were calculated as described above for each sample. Tables 8.1, 8.2, and 8.3 show the results of these calculations.

The introduction of pores in a material will reduce its effective thermal conductivity [19]. The results obtained in this work also support such an argument. The pore size could be at micron level or even finer, as generated at nanoscale by nanocasting [20]. Although, thermal conductivity of each sample was not measured directly in this experiment, the reduction in thermal conductivity can be observed by the increased temperature gradient for the most porous material. As shown in Tables 8.1, 8.2, and 8.3, the temperature gradients increase as the porosity of the material increases. It can be concluded that the most porous material has the least thermal conductivity and as a result the material conducts the least amount of heat, which contributes to a higher temperature gradient across the material. The image in Fig. 8.15 gives a close-up of a porous Bi-Sn sample, produced by the infiltration casting experiment.

Figure 8.15 Photograph of a porous Bi-Sn sample.

In this work, the temperature gradient was assumed to be constant during the time the voltage output was recorded for each sample. This assumption could contribute to possible percent errors since the heating element used had a fluctuating heating output. To minimize this error, the average voltage output for the first ten seconds of the experiment was used in the calculations.

The three samples used in this experiment varied in porosity levels from low to high. The samples used were also relatively short in length. The short length of the material contributes to a more uniform temperature distribution and a decrease in temperature difference as the time increases, which will result in a decrease in the voltage output of the material. This can be observed by studying the voltage output behavior as time increases. As shown in the plots of voltage versus time, the value of the voltage levels off as time increases. Having longer samples will ensure a higher temperature difference between the hot and cold ends of the material, which will increase the Seebeck coefficient and the figure of merit (efficiency).

Another source of error in this experiment could be the oxidation of our samples. Oxygen acts as an acceptor and can cause an inversion of the dominant carrier sign n- to p-type at the surface of the material [21]. Oxidation can also alter the thermal conductivity and electrical conductivity of the material. It can be seen that our material was slightly oxidized, since the surface of the metal was not shiny, which can be a source of error and a cause for lower figure of merit (efficiency). Since NaCl (salt) was used to create the pores in the samples, the formation of chloride on the surface and in the pours can accelerate the oxidation process. In addition, use of compressed air (which contains oxygen) to shoot the molten metal into the salt could be another factor contributing to the oxidation of our material. To minimize the oxidation due to the oxygen available in the air, compressed argon gas can be used in the experiment in lieu of compressed air.

8.5 Conclusion

In this chapter, we presented the infiltration casting approach to obtaining a porous Bi-Sn TE energy conversion material. The level of porosity in each sample varied from low to high. The samples were exposed to a temperature gradient, and their voltage outputs were recorded using a microsensor. The voltage output versus time was plotted for each sample at various temperature gradients and are shown. The Seebeck coefficient and the figure of merit value for each sample were also calculated. The results can be used to evaluate the TE energy conversion behavior of the porous material. The average Seebeck coefficient for the three samples, from least porous to most porous, are -1.64×10^{-5} (V/K), -2.55×10^{-5} (V/K), and -1.54×10^{-5} (V/K). The average ZT value for the three samples, from least porous to most porous, are 0.012, 0.031, and 0.010.

References

1. Freihaut, J. (2011). CHP technologies and applications, U.S. Department of Energy, Mid-Atlantic Clean Energy Application Center, from: http://energy.gov/sites/prod/files/2013/10/f4/fupwg_fall11_valentine.pdf.
2. Winkler, J., Aute, V., Yang, B., Radermacher, R. (2006). Potential benefits of thermoelectric elements used with air-cooled heat exchangers, International Refrigeration and Air Conditioning Conference, Purdue University, Paper No. 813.
3. Chung, Y. W. (2007). *Introduction to Materials Science and Engineering*, Boca Raton, FL: Taylor & Francis Group, Chapter 3.
4. Dresselhaus, M. S. (Fall 2004). Nano to Macro Transport Processes, MIT OpenCourseWare, Course Number: 2.57, Lecture 20.
5. The Fundamentals of Thermoelectrics, A Bachelor's Laboratory Practical, Nano Physics Group, Munchen, Germany.
6. Tian, Z., Lee, S., Chen, G. (2013). Heat transfer in thermoelectric materials and devices, *J. Heat Transfer*, **135**(6), pp. 1–11.
7. Velmre, E. (2007). Thomas Johann Seebeck (1770–1831), *Est. Acad. Sci. Eng.*, **13**(4), pp. 276–282.
8. Goldsmid, H. J. (2010). *Introduction to Thermoelectricity*, New York, NY: Springer.
9. Riffat, S. B., Ma, X. (2002). Thermoelectrics: a review of present and potential applications, *Appl. Therm. Eng.*, **23**(2003), pp. 913–935.
10. Cengel, M. A., Boles, M. A. (2008). *Thermodynamics: An Engineering Approach*, 6th ed., New York, NY: McGraw-Hill, Chapter 11.
11. Lundstrom, M. (2011). Thermoelectric effects: physical approach, Lecture 4, Near Equilibrium Transport: Fundamentals and Applications,

Purdue University, West Lafayette, Indiana, from: https://nanohub.org/resources/11747.

12. Incorpera, F. P., DeWitt, D. P., Bergman, T. L., Lavine, A. S. (2007). *Introduction to Heat Transfer*, 5th ed., Hoboken, NJ: John Wiley & Sons, Chapter 3.

13. Kim, H., Gibbs, Z., Tang, Y., Wang, H., Snyder, G. J. (2015). Characterization of Lorenz Number with Seebeck coefficient measurement, *APL Mater.*, **3**(4), pp. 1–5.

14. Snarskii, A. A., Zhenirovskii, M. I., Bezsudnov, I. V. (2011). Wiedemann-Franz law in thermoelectric composites, *J. Thermoelectr.*, **3**(2006), pp. 57–63.

15. Zide, J. M. O., Vanshee, D., Bian, Z. X., Zeng, G., Bowers, J. E., Shakouri, A., Gossard, A. C. (2006). Demonstration of electron filtering to increase the Seebeck coefficient in $In_{0.53}Ga_{0.47}As$/ $In_{0.53}Ga_{0.28}Al_{0.19}As$ superlattices, *Phys. Rev. B*, **74**(20), p. 205335.

16. Smith, D. S., Alzina, A., Bourret, J., Nait-Ali, B., Pennec, F., Tessier-Doyen, N. (2013). Thermal conductivity of porous materials, *J. Mater. Res.*, **28**(17), pp. 2260– 2272.

17. Callister, W. D., Rethwisch, D. G. (2014). *Material Science and Engineering, an Introduction*, 9th ed., Hoboken, NJ: John Wiley & Sons, Chapter 15.

18. Krzhizhanovskii, R. E., Sidorova, N. P., Bogdanova, I. A. (1974). Experimental investigation of the electrical resistivity of some molten bismuth-tin binary alloys and of the thermal conductivity of bismuth, tin, and eutectic bismuth-tin alloy, *J. Eng. Phys.*, **26**(1), pp. 33–36.

19. Miyazaki, K., Tanaka, S., Nagai, D. (2012). Heat conduction of a porous material, *J. Heat Transfer*, **134**(5), pp. 1–7.

20. Lu, A., Schuth, F. (2006). Nanocasting: a versatile strategy for creating nanostructured porous materials, *Adv. Mater.*, **2006**(18), pp. 1793–1805.

21. Rogacheva, E. I., Tavrina, T. V., Grigorov, S. N., Nashchekina, O. N., Volobuev, V. V., Fedorov, A. G., Nasedkin, K. A., Dresselhaus, M. S. (2002). Effects of oxidation on thermoelectric properties of PbSe thin films, *J. Electron. Mater.*, **31**(4), pp. 298–303.

Chapter 9

Thermoelectric Materials Made by Injection Casting

This chapter introduces recent research advances in manufacturing semiamorphous composite materials containing thermoelectric nanoscale grains by controlled injection casting and presents the results of a study on the thermoelectric behavior of nanocomposite materials. The research results show the existence of nanoscale grains in amorphous metallic substrates. It is concluded that semiamorphous nanocomposites may decouple thermal conductivity and electrical conductivity, thus improving the thermoelectric energy conversion efficiency of nanocomposite materials.

9.1 Introduction

Thermoelectric (TE) conversion materials are capable of clean energy generation from geothermal, solar, and other types of waste heat. However, the inherent conflicts among high Seebeck coefficient, high electrical conductivity, and low thermal conductivity lead to low values of TE figure of merit (ZT), hindering their application in high-efficiency energy harvesting. It is necessary to explore high-efficiency TE energy conversion composite materials. Through controlled injection casting, an amorphous metal matrix containing Bi-Te alloy nanoscale grains as the high TE power generating

Nanomaterials for Thermoelectric Devices
Yong X. Gan

ISBN 978-981-4774-98-7 (Hardcover), 978-0-429-48872-6 (eBook)
www.panstanford.com

domains with tailored electrical and thermal transport behavior was made. This type of amorphous metal matrix nanocomposite material is expected to resolve the inherent property conflicts and help acquire high *ZT* values for efficient TE energy conversion. This hypothesis is built upon the theoretical analysis of the Boltzmann transport equation and preliminary TE measurements, that is nanocomposites can enhance the quantum confinement and generate energy-filtering effect, allowing a large Seebeck coefficient and high electrical conductivity together for a large TE power factor. Bi-Te nanoscale grains encapsulated in an amorphous matrix with a very high interface area can scatter phonons and disrupt heat conduction, resulting in high electrical conductivity but low thermal conductivity. The effect of manufacturing parameters on the morphology of the nanocomposite were investigated to the tune the nanostructures for a high *ZT*. The chapter deals with fundamental research and TE energy conversion application.

The first part of this chapter is on designing and setting up a new manufacturing machine with the capability of performing controlled injection casting under the actions of high-pressure argon gas and an external magnetic field. The machine was built on the basis of a levitation melting system. The induction coil of the system allows one to melt metal elements. Self-designed molds with nanoscale channels or pillars were used to generate a high cooling rate to facilitate the amorphous composite material formation.

The second part of this chapter is on understanding the solidification process and the nanoscale grain formation mechanism of the levitating melted alloys in nanometer channel cavities during the injection casting. Specifically, the following fundamental questions should be answered: How fast do the molten alloys solidify? At which cooling rate level do the alloys solidify into semiamorphous composite materials? What is the role of the magnetic field in the formation of the nanoscale grain? By modeling the heat transfer and the momentum transfer in the injection casting, the cooling rate will be evaluated. By performing a magnetohydrodynamic (MHD) analysis, we will find out how effectively a constant magnetic field suppresses the grain growth and prevents the mixing of the chemical composition when casting complex alloys. Such studies could further our understanding of the science underpinning the new manufacturing process during which magnetic and mechanical actions coexist.

The third part of the chapter is on the structure assessment of the composite materials. Optical and scanning electron microscopic (SEM) studies of the microstructure were performed to reveal the morphology of the composites. The grain size of Bi-Te was measured by microscopic studies.

The fourth part of the chapter is on characterization of the TE property of the composite materials. Experiments on Seebeck coefficient measurement were performed, and the results are presented. Fundamental studies on thermal and electric transfer behaviors were reviewed, and guidelines for exploring high-performance TE materials with multiple components and complex structures are discussed. The approach to an amorphous matrix holding TE Bi-Te alloy nanoscale grains is used. Dispersion of the nanoscale grains in the matrix can be achieved via magnetic field–assisted injection casting. Nanoscale grains are used to enhance the thermoelectricity, and specifically to increase the Seebeck coefficient. Such nanocomposites have complex structures and high interface areas for phonon scattering, leading to good TE properties of the composites, that is a high Seebeck coefficient, high electrical conductivity, and low thermal conductivity. Interplaying on structure analysis and TE property measurement were carried out to understand the influence of magnetic field and the mechanical force coming from the high-pressure argon gas on the structure and TE property of the injection-cast nanocomposite materials.

9.2 Brief Review of Injection-Casting Thermoelectric Nanocomposites

The proposed work focuses on controlled injection casting of nanocomposites containing amorphous matrices and nanoscale TE grains as high-performance energy conversion materials. Although the injection casting concept has recently been studied for various materials systems, including metallic alloys, ceramics, and composite materials [1–10], the technique has never been used for manufacturing complex TE energy conversion nanocomposites. Early studies show that magnetic field–assisted casting is effective in cladding steel slabs with different chemical compositions [11–15]. By extending the magnetic force–assisted injection casting process concept to complex semiamorphous TE composite material systems, it is possible to manufacture a nanocomposite with an amorphous matrix and dispersed nanoscale TE material grains. The challenges that still remain are as follows: how to use a single-step injection casting process to manufacture large-scale nanocomposites; how to increase the production rate of the nanostructures while keeping the cost associated with the composite manufacturing reasonable; and most importantly, how to manufacture nanocomposites with controlled architectures for functional applications, such as TE energy conversion. The scope of the proposed work is to advance amorphous-crystalline composite casting while ensuring high process yield, process and product repeatability and reproducibility, and optimized quality control, resulting in a commercially viable, scalable manufacturing process that can produce functional nanocomposites for high-efficiency TE energy conversion.

TE energy conversion allows the generation of electricity from various heat sources, as reported recently [16–20]. The energy sources could be solar irradiation, automotive exhaust heat, and waste heat from industrial processes [21–24]. Thermoelectricity is renewable, reliable, and scalable [25–27]. It is expected to play an increasingly important role in meeting the energy challenges in the future. The efficiency of TE energy conversion depends on the figure of merit of the TE material used [21–27]. The value of ZT is proportional to the electrical conductivity σ, the absolute temperature T, and the square of the Seebeck coefficient S, but inversely proportional to the thermal conductivity κ [25–27]. Obviously, materials with a large S, a high σ, and a low κ are required to maximize ZT for high-efficiency TE conversion. However, there are obvious conflicts among these parameters in regular materials that are a challenge to resolve. Usually, reducing the carrier concentration of a material enhances its S but decreases its σ. Increasing the carrier concentration raises σ and κ simultaneously [28–31]. Therefore, regular materials cannot solve the inherent conflicts among S, σ, and κ, resulting in low efficiency of TE energy conversion. The potential application of TE conversion is significantly hindered by low ZT values of currently available materials. High-ZT materials are sought to make TE conversion cost effective and competitive so that it may provide a feasible solution to the worldwide energy issue. For decades, many attempts have been made to resolve these conflicts [32–41]. Although most of the early work was focused on inorganic materials [32–40, 42–43], organic materials have recently received great attention as TE materials because of their tunable TE properties [44–47]. Organic TE materials have some attractive features for utilization. They are lightweight and flexible. They demonstrate higher power density even though they show less conversion efficiency than inorganic materials; and they are abundantly available and easy to process, easy to incorporate into devices, reusable, and environmentally benign. The limitation of organic TE materials lies in their poor thermal stability. By contrast, inorganic TE materials such as chalcogenide metallic compounds exhibit high temperature resistance and their ZT values could be as high as 1.0 [32–40]. The disadvantages are that they are high in cost due to very limited reserves and the main elements are hazardous and expensive heavy or rare earth metals; they usually require a complicated manufacturing process, which consumes huge quantities of energy; they are also very difficult to incorporate into TE devices and difficult to separate and recycle. Therefore, combining different materials to form composites will make TE energy a cost-effective and feasible alternative energy source if they can demonstrate ZT values similar to or better than those of the currently available materials.

To explore materials suitable for the matrices of TE composite materials, it is worthy to compare the thermal and electric conductive behaviors of different categories of materials. Organic materials are intrinsically poor thermal conductors, which makes them ideal for reducing the κ values of TE composite materials [39, 40]. Generally, organic materials also show very low electrical conductivity and this excludes them as feasible candidates for the matrices of TE composite materials. Composite technology has made it possible to turn organic materials into semiconductors or metal-like materials by synthesizing conjugated structures or incorporating conductive fillers [41–48]. Although there are conducting polymers available [49–55], the TE properties of such organic materials can't survive under elevated temperature conditions.

Oxides and various other forms of ceramics as TE materials have been extensively studied due to their good thermal stability in a corrosive environment and at high temperatures [56–82]. There is no work done on chalcogenide metallic compounds-amorphous metal multicomponent composite materials for TE energy conversions. In this proposal, a novel semiamorphous nanocomposite will be investigated. It consists of a chalcogenide metallic compound (Bi-Te,) with a strong Seebeck effect, and an amorphous metal alloy as the substrate. Bi-Te nanoscale grains encapsulated in an amorphous matrix will be manufactured by controlled injection casting into molds with nanometer cavities to generate a high cooling rate. By applying a level magnetic field, we can suppress the growth of the Bi-Te grains during the solidification process. This allows the amorphous substrate to properly accommodate the active TE component, the Bi-Te alloy. The structure and TE property of the nanocomposite will be studied as well.

We have performed research work on manufacturing a bismuth-based alloy with a porous structure via induction heating followed by injection casting, developing a fundamental understanding of the transport processes related to TE energy conversion [83]. The casting manufacturing process can produce a bismuth-based porous material. Induction heating was applied to melt the Bi-Sn alloy in a quartz tube, and the molten alloy was cast into loosely compacted sodium chloride powder. After the sodium chloride powder was dissolved using water, pores were generated in the Bi-Sn alloy. The true porosity of the alloy can be controlled to as high as 60% in volume. The TE property of the material has been studied to explore the application of this material for energy conversion. The experimental results show that the Seebeck coefficient of the porous bismuth material is independent of porosity. The porosity of the material can be controlled through manufacturing parameters. The higher the porosity, the slower is the heat conduction in the material.

Our early work [83] provides experience in casting a Bi-Sn alloy into a porous template made of packed salt particles. On the basis of preliminary studies, we have extended the research work to composite material systems to manufacture amorphous matrix composites by increasing the cooing rate. Up to now, there is little research work done on casting high-temperature TE composite materials under both mechanical and magnetic actions. The challenges of this new manufacturing technology include the following: how to use a single-step process to manufacture large-scale composite materials; how to increase the production rate of the nanostructures while keeping the cost associated with the manufacturing process reasonable; and most importantly, how to manufacture special composites with controlled nanoscale grains and amorphous matrices for use in TE energy conversion. The focus of this chapter is on understanding the fundamentals of magnetic force–assisted injection casting manufacturing and to use this new approach to make nanoscale grain–amorphous matrix composites. A commercially viable, scalable manufacturing process ensuring high process yield, process and product repeatability and reproducibility, and optimized quality control can be obtained for producing functional nanocomposites for high-efficiency TE energy conversion.

9.3 Manufacturing Process and Characterization

A nanoscale template mold was first made from anodic aluminum oxide (Al_2O_3, AAO) and a polycarbonate (PC). Then, the magnetic force–assisted injection casting machine was designed and made on the basis of a levitation melting machine. The TE nanocomposites were made using the injection casting machine. The Seebeck coefficient of the composites was measured using the CHI 440C workstation. Electrical conductivity maps of the composites were generated using an atomic force microscope (AFM). The scope of the manufacturing research and characterization is given in Fig. 9.1.

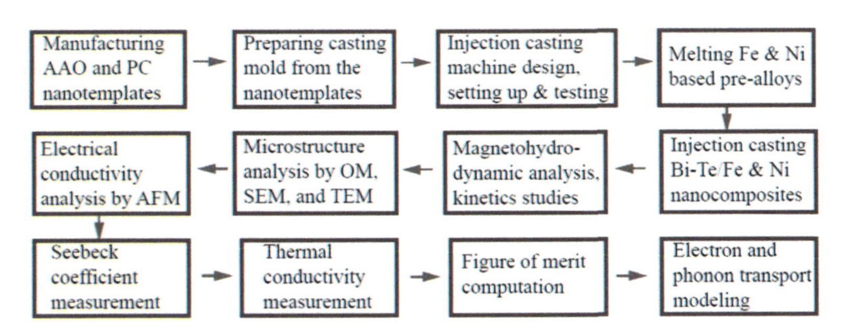

Figure 9.1 Manufacturing and characterization flowchart.

9.3.1 Nanoscale Template Mold Manufacturing

First, an AAO template with a pore diameter of 200 nm was made. The two-step anodic oxidization method [84, 85] was used to obtain AAO with a uniform size and a thin barrier layer. High-purity Al sheets were anodized using a regulated DC power supply in oxalic acid solution. Before they were anodized, the Al plates were degreased in trichloroethylene, followed by ultrasonic cleaning in acetone. Then the samples were rinsed with methanol and distilled water separately. After that, the Al plates were etched in NaOH. The Al was electropolished in a HNO_3-methanol solution. After that, anodizing was performed in $H_2C_2O_4$. The Al served as the anode and a Pt foil or a graphite plate the cathode. After the first anodizing, a strip-off process was carried out in H_3PO_4. The exposed and well-ordered concave patterns on the Al substrate acted as a self-assembled mask for the second anodizing. After the second anodization, AAO templates with uniform nanopores were obtained, as schematically shown in Fig. 9.2a. The depth of the nanoscale pores can be controlled by the oxidizing time. Polymer impregnation (Fig. 9.2b) was done, followed by physical vapor deposition of copper (Fig. 9.2c). After dissolving the Al and AAO using NaOH (see Fig. 9.2d), electrodeposition of Cu was performed (Fig. 9.2e). The final product, in the form of a nanoporous metal template (Fig. 9.2f), was obtained by removing the polymer using an organic solvent. A SEM image of the nanoporous template mold is shown in Fig. 9.2g. This nanoporous copper metal template was used later for controlled injection casting.

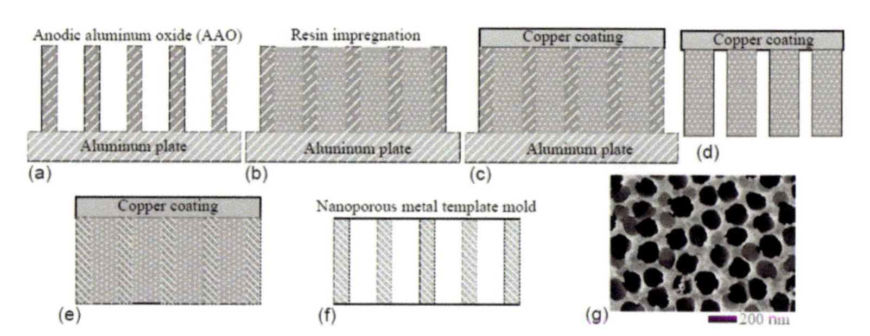

Figure 9.2 Injection casting mold made from anodic aluminum oxide (AAO): (a) AAO, (b) polymer impregnation, (c) Cu deposition, (d) Al and AAO removal, (e) Cu electroplating, (f) polymer removal, and (g) SEM image showing a typical copper nanotemplate mold with pores 200 nm in size.

The injection casting mold was also manufactured from a PC porous template. A diblock copolymer of PC-polymethyl methacrylate (PMMA) was made in a high-voltage DC electric field to allow PMMA to organize as microcylinders in a PC matrix. Then, porous PC was prepared through intensive ultraviolet (UV) exposure of PC-PMMA. In UV light, PMMA degraded and was removed from the PC matrix by subsequent chemical etching processes. A porous PC template as schematically shown in Fig. 9.3a and with the SEM image of Fig. 9.3b was filled with Cu by electrodeposition, as shown in Fig. 9.3c. After removal of the PC polymer by an organic solvent, a new mold comprising micropillars as shown in Fig. 9.3d was made for injection casting. The diameter of the pillar is about 4 microns.

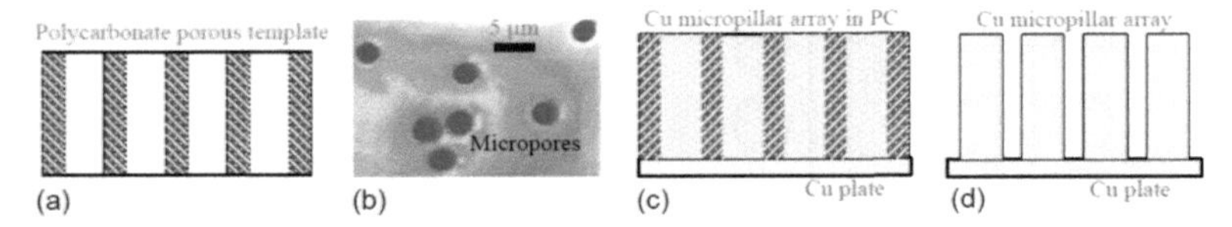

Figure 9.3 Casting mold made from PC: (a) PC, (b) SEM image, (c) Cu deposition, and (d) PC removal.

9.3.2 Controlled Injection Casting Machine Design

Studies of injection casting a Bi-based alloy into porous media have been performed. Figure 9.4a schematically shows how the injection casting machine works. Figure 9.4b presents the equipment setup for the injection casting experiment. Vacuum induction melting followed by injection casting to make the porous Bi-Sn material was performed without using the nanoporous template. The quartz tube was half filled with NaCl. Pieces of eutectic Bi-Sn alloy were placed above the salt. A valve tee connection was used to connect the quartz tube, the compressive gas, and the vacuum equipment. The molten alloy was injected into the salt powder by the following procedure. The first step was to create vacuum in the salt. The compressor valve was closed and the vacuum valve was opened prior to using the heat inductor. The quartz tube was then placed in the induction coils. The electromagnetic field is present for induction heating to work. After the alloy was melted, the compressor valve was opened to place pressure on the molten metal and cast the metal into the loosely compacted salt. After the molten alloy solidified, water was used to dissolve the salt from the cast. The porous metal sample shown in Fig. 9.4c was obtained, and a SEM image was taken as shown in Fig. 9.4d.

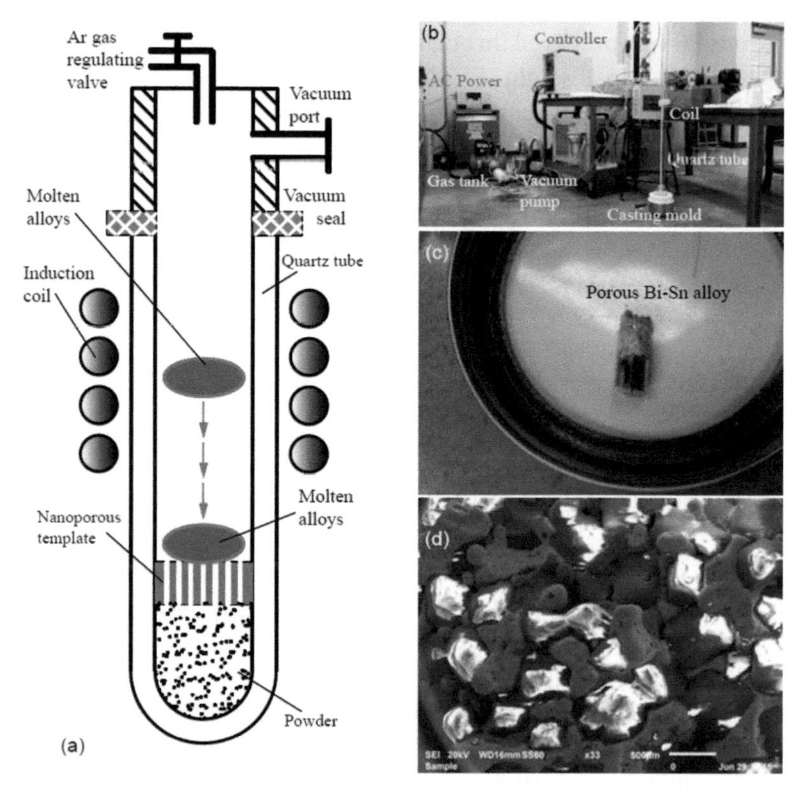

Figure 9.4 Controlled injection casting machine and the manufacturing product: (a) schematic of the casting machine, (b) the setup of the machine, (c) casting product, and (d) SEM image of the alloy.

Our study [83] also showed that the microstructure of the Bi-Sn alloy is cooling rate sensitive. If it is cooled down slowly in air, a well-developed eutectic structure is presented, which is revealed by Fig. 9.5a. Rapid cooling in a steel mold leads to the formation of bismuth-rich nanoparticles separated by Sn-rich laminars, as shown in Fig. 9.5b. In recent work, a modified injection casting design as shown in Fig. 9.6 was performed. Nanotemplates were placed in a water-chilled Cu mold to ensure an increased cooling rate for amorphous substrate formation. Magnets were put surrounding the molten jet to examine the magnetic field–suppressed Bi-Te grain growth.

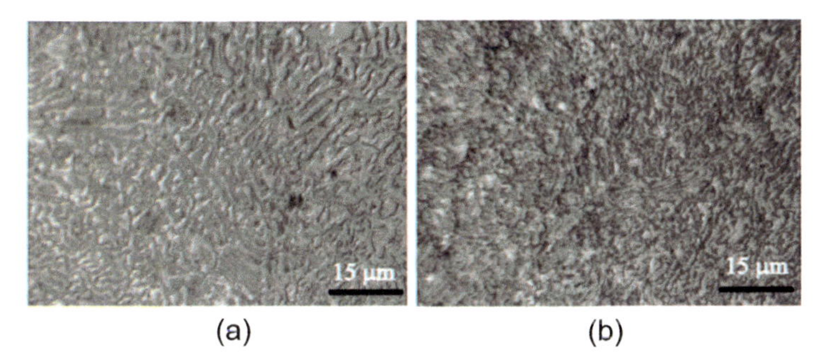

Figure 9.5 Microstructure of injection cast Bi-Sn: (a) cooling in air and (b) rapid cooling in steel mold.

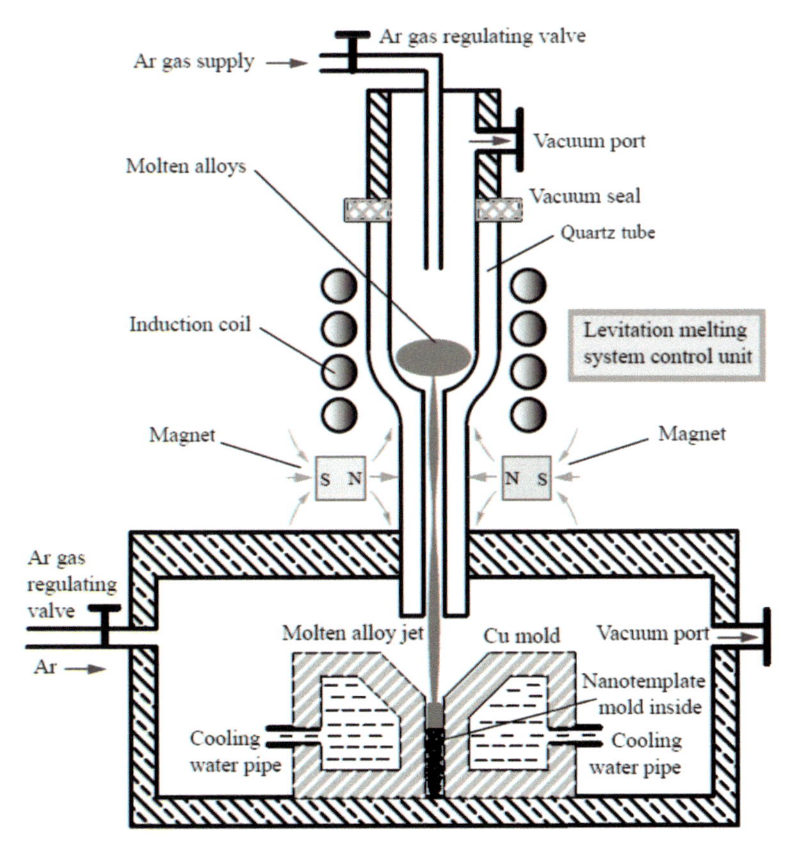

Figure 9.6 Modified injection casting machine with an increased cooling rate due to installation of the water-chilling copper mold and the magnetic field–assisted grain growth suppression function.

9.3.3 Amorphous Matrix Alloy Design, Prealloy Melting, and Full-Alloy Preparation

To determine a suitable amorphous matrix for holding the active TE Bi_2Te_3 nanocrystals, experimental studies on prealloy melting were done. Recent research on amorphous reveals that Zr-, Ti-, Cu-, or Fe-based alloys are relatively easy to vitrify [86–106]. Cu-Zr-Al nanocrystal–containing amorphous composites can be synthesized by the copper-mold suction casting method. It is found that the existence of nanocrystals dispersed homogeneously in the glassy matrix can improve the properties, for example, by increasing the corrosion resistance of a $Cu_{47.5}Zr_{47.5}Al_5$ alloy [90]. Iron-based glassy alloys are attracting more and more attention due to the high strength, high glass transition temperature, and good magnetic property [92, 94, 98, 105]. In this study, we select an Fe-Ni alloy with some glass forming elements, such as P, B, Cr, Mo, and C, because of the strong response of the alloy system in electromagnetic fields. Our recent studies show that such an alloy melts fast in induction heating. Preliminary studies by another group [83] have proved that in a "level DC magnetic field," the magnetic flux leads to the separation of the chemical composition or prevents mixing. On the basis of previous knowledge, by adding the magnet into the manufacturing machine as shown in Fig. 9.6, we can separate the nanocrystal Bi-Te and the amorphous Fe-Ni alloy. A laminar structure like that in Fig. 9.5 can be obtained.

The following steps were used for prealloy preparation. First, the elements were weighed, including Fe, Mo, Cr, Ni, B, and C. Then, the B and C were crushed together in a steel cylinder tool into fine powders. The six elements were loaded into a quartz tube, and the quartz tube was vacuumed. The elements were melted in vacuum using the induction heating system as shown in Fig. 9.4b. After the prealloy was made, full-alloy preparation was conducted. We weighed the element P and added 10% extra considering the evaporation loss of the element.

The following steps are adopted to prepare the full alloy. Crush the prealloy into small pieces. Load P at the bottom of a quartz tube. Then vacuum the quartz tube and purge Ar with partial pressure and start melting experiment by increasing the power slowly. Since P has a very low melting point. The power level will be set low first.

Then, increase the power slowly to allow the prealloy to melt and mix well with P. Refluxing the full alloy with boron oxide is done to homogenize the distribution of the elements and eliminate oxygen contamination. To do the refluxing, weigh B_2O_3 so that the ratio of boron oxide to the alloy by weight is about 1:3. Set the boron oxide pieces at the bottom of a quartz tube. Melt the mixture of boron oxide and the full alloy in vacuum. Wait for the boron oxide bubbling for 15 min. The full alloy is obtained.

9.3.4 Controlled Injection Casting of Nanocomposites

The injection casting process is considered as the key part of making nanocomposite materials with Bi-Te nanocrystals in an Fe-based amorphous matrix. A bigger quartz tube, with an outside diameter of 15 mm and an inside diameter of 10 mm, was connected with a smaller quartz tube, with an inside diameter of 4 mm, by necking down the bigger tube in a propene oxygen torch. Then, the Fe-based full alloy and preweighed Bi-Te alloy were loaded into the bigger quartz tube as shown in Fig. 9.6. Part of the smaller quartz tube was inserted into the vacuum chamber. The quartz tube was vacuumed. Then, the vacuum valve was closed. The two alloys were heated at a sufficient power level. Once the two alloys melted, the melt was exposed to high-pressure Ar, which allowed the molten alloys to fill the template located in the copper mold chilled by icy water. Nanocomposites containing an amorphous Fe-Ni matrix and Bi-Te alloy nanocrystals formed.

9.3.5 Manufacturing Process Analysis

One of the fundamental problems associated with the manufacturing process is how fast the molten jet as shown in Fig. 9.6 is injected into the porous template under the combined magnetic and mechanical forces. Assume that the velocity of the jet is $\boldsymbol{V}$, the argon pressure used for injection casting is p, and the Lorentz force is $\boldsymbol{F}_{\mathrm{m}}$. The motion of the molten jet can be expressed by the continuity equation, $\nabla \bullet \vec{V} = 0$ and the following momentum equations [107].

$$\frac{\partial \vec{V}}{\partial t} + \left(\vec{V} \bullet \nabla \right) \vec{V} = -\frac{1}{\rho} \nabla p + v \left(\nabla^2 \vec{V} \right) + \vec{g} + \frac{1}{\rho} \vec{F}_{\mathrm{m}}, \qquad (9.1)$$

where ρ is the density and ν is the viscosity of the liquid jet. Considering the magnetic field effect, Maxwell's equations and Ohm's law are used to correlate the electric and magnetic field density and the velocity of the injected molten metals [108–112]:

$$\nabla \times \vec{E} = -\frac{\partial \vec{B}}{\partial t} = -\mu \frac{\partial \vec{H}}{\partial t}, \tag{9.2a}$$

$$\nabla \bullet \vec{B} = 0, \tag{9.2b}$$

$$\nabla \times \vec{H} = \vec{J}, \tag{9.2c}$$

and

$$\vec{J} = \sigma \left(\vec{E} + \vec{V} \times \vec{B} \right), \tag{9.2d}$$

where σ is the conductivity and μ is the permeability of the melt. ***H*** is the magnetic field strength. ***B*** is the magnetic flux density. ***E*** is the electric field, and ***J*** is the electric current density. If we define the scalar potential function φ [113], the electric field intensity can be expressed by the potential gradient, that is $\vec{E} = -\nabla \phi$. The current satisfies the continuity $\nabla \bullet \vec{J} = 0$.

To solve the couple problem as expressed by Eqs. 9.1 and 9.2, we will use the numerical method. COMSOL multiphysics software will be used to compute the velocity field ($\boldsymbol{V}$), the Lorentz force ($\boldsymbol{F}_{\mathrm{m}}$), and the potential function (φ). The solution for $\boldsymbol{V}$ will be further applied for mass transfer analysis to determine the distribution of Bi using the diffusion equation.

$$\frac{1}{D}\left(\frac{\partial C}{\partial t} + \vec{V} \bullet \nabla C \right) = \nabla^2 C, \tag{9.3}$$

where D is the diffusion constant of Bi in the melt and C is the concentration of Bi. We only take Bi as the candidate element for analysis because the distribution of Bi allows us to predict the nanocrystal formation of the active TE compound, Bi_2Te_3. The results from this analysis were compared with those from the transmission electron microscopic (TEM) observation, which provides us the proof for validating nanoscale grain formation in the injection casting process.

9.3.6 Microstructure and Thermoelectric Property Characterization

The microstructure of the composites was studied. X-ray diffraction was conducted to determine the amorphous state of the matrix. TEM and scanning transmission electron microscopy (STEM) analysis were performed to determine the crystal structure of the Bi-based nanoscale grains. The dislocation density within the nanoscale grains was calculated to reveal the interaction between the amorphous matrix and the crystalline Bi-Te nanoscale grain. Seebeck coefficients were measured. Figure 9.7a is a schematic drawing to show measurement of the Seebeck coefficients. The experimental setup is given in Fig. 9.7b.

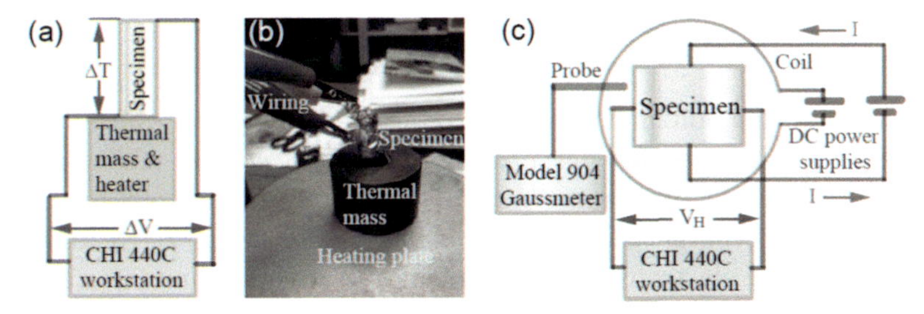

Figure 9.7 Illustrations and picture showing (a) Seebeck coefficient measurement, (b) experimental setup for measuring the Seebeck coefficient, (c) and an illustration for measuring the Hall effect coefficient.

During the Seebeck coefficient measurement, the two ends of a sample were bonded to two separate wires with an Ag-based conductive adhesive, which allows high electrical conductivity at the wire/sample interface. Then, one end of the specimens as the hot end was heated up by the heater to a specific temperature and the other end, as the cold end, was kept in air at the ambient temperature of 25°C. The hot-end temperature ranges from 40°C to 450°C. The Seebeck coefficient was calculated by the ratio of $\Delta V/\Delta T$. ΔV represents the voltage difference between the hot and cold ends. ΔT is the temperature difference between the two ends. The quantum Hall effect was also investigated to understand the size-dependent TE behavior. A Hall effect measurement apparatus as sketched in Fig. 9.7c was used to measure the Hall voltage V_H. The

Hall effect measurement results were used to determine the charge carrier density and the electron mobility, which allow us to study the electron transport behavior of the nanocomposites.

Theoretical studies for establishing the necessary Seebeck coefficient model were conducted to validate that the TE property of the composites can be enhanced by the formation of nanocrystals. As the start point, the Cutler–Mott formula [114], which was developed for degenerated semiconductors, was used to estimate the Seebeck coefficient of the composite materials containing nanocrystals of different sizes within the amorphous matrix, S, that is

$$S = -\frac{\pi^2}{3e}k^2T\frac{\partial \ln[\sigma(E)]}{\partial E}\bigg|_{E=E_\mathrm{F}} \approx -\frac{\pi^2}{3e}k^2T\left(\frac{\partial \ln[\sigma(n)]}{\partial E}\bigg|_{E=E_\mathrm{F}} + \frac{\partial \ln[\sigma(\tau)]}{\partial E}\bigg|_{E=E_\mathrm{F}}\right), \tag{9.4}$$

where k is the Boltzmann constant, σ the conductivity, e the absolute value of the electron charge, n the density of charge carriers, τ the scattering time, and E_F the Fermi energy. Since Bi_2Te_3 is n-type, the charge carriers in the Bi-Te nanocrystal/Fe-Ni alloy interface are electrons. The carrier concentration per unit interface area can be determined approximately by

$$n \approx \int_0^{E_F} \frac{4\pi}{h^2}m_\mathrm{eff}\left(\sum_\nu \Phi(E-E_\nu)\right)dE = \frac{4\pi}{h^2}m_\mathrm{eff}\left(\sum_\nu (E-E_\nu)\Phi(E-E_\nu)\right), \tag{9.5}$$

where ν is the index of the discrete energy levels, Φ the Heaviside step function whose value is zero for a negative argument and equals 1 for a positive argument, m_eff the effective mass of the charge carriers, and E_ν the νth energy level. E_ν can be determined by $E_\nu = -\frac{h^2\nu^2}{8dm_\mathrm{eff}}$, where d is the nanocrystal size. From the above analysis, it is clear that in order to increase the Seebeck coefficient we should increase both conductivity σ and scattering time τ. The smaller the size of the Bi-Te nanocrystals, the more negative is the value of E_ν. Therefore, it is necessary to make very fine nanocrystals because the finer the nanocrystals, the bigger is the interface area, and the longer is the scattering time, in addition to a more negative value of E_v. Our current task focuses on developing a new model specifically for nanocrystals encapsulated in amorphous matrices. By tuning the size of the nanocrystals through a controlled cooling rate in the injection casting, the conductivity σ, the scattering time τ, and the vth energy level E_v can be calculated and compared with experimental measurements. On the basis of the comparative studies, necessary modification of the model is underway.

9.3.7 Electrical and Thermal Conductivity Measurement

Electrical and thermal conductive properties were characterized for evaluating the energy conversion effectiveness of the nanocomposites. A measuring system containing the TD-8561 thermal conductivity apparatus was used to determine the thermal conductivity of the nanocomposites (k). A GLX Xplorer data acquisition unit connected to an I-V sensor and two temperature sensors were used to measure the electrical conductivity (σ) and to calculate the thermopower (U). Electrical conductivity maps at micro- and/or nanoscale were obtained using the Innova conductive AFM purchased from Bruker. The electrical conductivity mapping results were used to examine the size-dependent transport behavior of the composite materials. On the basis of both thermal and electrical measurements, the TE figure of merit, ZT, was calculated using the formula $ZT = (\sigma U^2T)/k$. The TE energy conversion efficiency will be calculated from the figure of merit data. Varying the Bi-Te nanocrystal size causes changes in the electrical conductivity and thermal conductivity of the nanocomposite materials. The effect of the Bi-Te amount on conductivity was studied as well.

9.3.8 Modeling Phonon Damping in Injection-Cast Nanocomposites

Phonon transport can be considered as wave propagation. Multiscale approaches were used to model phonon transmission. First, considering the phonon waves in perpendicular incidence to the lamellar nanocomposites (nanocrystals separated by amorphous lamellar), each layer can be modeled as an individual transmission line. The following Riccati equation [115] can be established to describe the decaying behavior of phonons:

$$\frac{d\Gamma}{dx}-\frac{4i\pi}{\lambda}n(x)\Gamma+\frac{1}{2Z}\frac{dZ}{dx}(1-\Gamma^2)=0, \tag{9.6}$$

where Γ is the reflectivity of phonon, x is the thickness of the material, $n(x)$ is the equivalent refractivity of the nanocomposites at the thickness of x, λ is the wavelength of the phonon, Z is the impedance of the materials, and i is the unit of imaginary numbers.

Defining two parameters $\alpha=\frac{4i\pi}{\lambda}n(x)$ and $\beta=\frac{1}{2Z}\frac{dZ}{dx}$ and assuming that a is constant for a fixed wavelength of the phonon wave, the following relation holds at the equilibrium point:

$$\frac{d\Gamma}{dx}=b\Gamma^2+\alpha\Gamma-\beta=0, \tag{9.7a}$$

and

$$\Gamma=\frac{-\alpha\pm\sqrt{\alpha^2+4\beta^2}}{2\beta}. \tag{9.7b}$$

The determinant of the linear matrix of Eq. 9.7a or 9.7b at the equilibrium point, $A(\beta)$, is:

$$A(\beta)=\frac{\partial}{\partial\Gamma}\left(\frac{d\Gamma}{dx}\right)=\frac{\partial}{\partial\Gamma}(\beta\Gamma^2+\alpha\Gamma-\beta)=2\beta\Gamma+\alpha \tag{9.8}$$

To solve the Eigen value, η, let I be the unity matrix. From $\left|A(\beta)-\eta I\right|_{\frac{\partial\Gamma}{\partial x}=0}=0$, the Eigen value for $A(\beta)$ can be determined to be $\eta=\pm\sqrt{\alpha^2+\beta^2}$. By setting η = 0, we can find the condition under which the static bifurcation of Eq. 9.6 occurs. That is $\beta=\pm\frac{\alpha}{2}i$; that is $\Gamma=\pm i$ or $|\Gamma|=1$. Since the mode of Γ corresponds to the absolute value of the reflectivity of phonon, this result predicts that a nanocomposite could be a heat sink. Although the mathematical treatment described here is just for a simple case to demonstrate the idea, to address phonon transport in more complex cases, networks consisting of many transmission lines should be used to do numerical analysis. For the nanoscale analysis, phonon transport can be quantized as lattice vibrations. A vibration Schrodinger equation for a periodic lattice can be established. A solution to this equation can be found both analytically and numerically.

Such analysis can be used to address the question of why thermal conductivity can be reduced in nanocomposite materials. It is also necessary to study the effects of manufacturing conditions on transport behaviors of composite materials.

9.4 Conclusions

Semiamorphous composite materials containing TE nanoscale grains can be manufactured by controlled injection casting. We obtained some preliminary results of our study on the TE behavior of nanocomposite materials. The existence of nanoscale grains in amorphous metallic substrates is demonstrated. Semiamorphous nanocomposites may be the solution for decoupling thermal conductivity and electrical conductivity and improving the TE energy conversion efficiency of devices made of nanocomposite materials.

References

1. Zhang, L., Li, D., Chen, X., Qu, X. H., Qin, M. L., He, X. B., Li, Z., Wang, J. Y. (2015). Microstructure and mechanical properties of MIM213 superalloy, *Mater. Chem. Phys.*, **168**, pp. 18–26.
2. Ahmad, Z., Yan, M., Tao, S., Liu, Z. (2013). $Fe_{69}B_{20.2}Nd_{4.2}Nb_{3.3}Y_{2.5}Zr_{0.8}$ magnets produced by injection casting, *J. Magn. Magn. Mater.*, **332**, pp. 1–5.
3. Liu, Z. Q., Qi, F. S., Li, B. K., Jiang, M. F. (2014). Vortex flow pattern in a slab continuous casting mold with argon gas injection, *J. Iron Steel Res. Int.*, **21**, pp. 1081–1089.
4. Zhang, B., Ye, F., Gao, Y., Liu, Q., Liu, S. C., Liu, L. (2015). Dielectric properties of $BADCy/Ni_{0.5}Ti_{0.5}NbO_4$ composites with novel structure fabricated by freeze casting combined with vacuum assisted infiltration process, *Compos. Sci. Technol.*, **119**, pp. 75–84.
5. Zhou, B. W., Yin, S. J., Tang, R. H., Yang, H. S., Ya, B., Jiang, B. Y., Fang, Y., Zhang, X. G. (2016). Study on fabrication of bulk metallic glassy composites by horizontal continuous casting method, *J. Alloys Compd.*, **660**, pp. 39–43.
6. Liu, Z. Q., Qi, F. S., Li, B. K., Cheung, S. C. P. (2016). Modeling of bubble behaviors and size distribution in a slab continuous casting mold, *Int. J. Multiphase Flow*, **79**, pp. 190–201.

7. Xu, T., Wang, C. A. (2016). Control of pore size and wall thickness of 3-1 type porous PZT ceramics during freeze-casting process, *Mater. Des.*, **91**, pp. 242–247.
8. Yang, M., Gong, Y., Yu, X., Feng, L., Shi, Y., Huang, Z., Xiang, X., Wei, J., Lu, T. (2016). Fabrication of Li_4SiO_4 ceramic pebbles with uniform grain size and high mechanical strength by gel-casting, *Ceram. Int.*, **42**, pp. 2180–2185.
9. Sheng, L. Y., Yang, F., Xi, T. F., Zheng, Y. F., Guo, J. T. (2013). Microstructure and room temperature mechanical properties of NiAl-Cr(Mo)-(Hf, Dy) hypoeutectic alloy prepared by injection casting, *Trans. Nonferrous Met. Soc. China*, **23**, pp. 983–990.
10. Zhang, L., Li, D., Chen, X., Yang, F. B., Wang, J. Y., Chen, B., Ma, X., Qu, X. H. (2015). Preparation of MIM213 turbine wheel with hollow internal structure, *Mater. Des.*, **86**, pp. 474–481.
11. Harada, H., Takeuchi, E., Zeze, M., Tanaka, H. (1998). MHD analysis in hydromagnetic casting process of clad steel slabs, *Appl. Math. Modell.*, **22**, pp. 873–882.
12. Zeze, M., Harada, H., Takeuchi, E., Ishii, T. (1993). Application of a DC magnetic field for the control of flow in the continuous casting strand, *Ironmaking Steelmaking*, **20**, pp. 53–57.
13. Takeuchi, E. (1995). Applying MHD technology to the continuous casting of steel slab, *JOM*, **47**, pp. 42–45.
14. Takeuchi, E., Zeze, M., Tanaka, H., Harada, H., Mizoguchi, S. (1997). Novel continuous casting process for clad steel slabs with level dc magnetic field, *Ironmaking Steelmaking*, **24**, pp. 257–263.
15. Yamamura, H., Toh, T., Harada, H., Takeuchi, E., Ishii, T. (2001). Optimum magnetic flux density in quality control of casts with level DC magnetic field in continuous casting mold, *ISIJ Int.*, **41**, pp. 1229–1235.
16. Whitney, R. S., Sánchez, R., Haupt, F., Splettstoesse, J. (2016). Thermoelectricity without absorbing energy from the heat sources, *Physica E*, **75**, pp. 257–265.
17. Hashim, H., Bomphrey, J., Min, G. (2016). Model for geometry optimization of thermoelectric devices in a hybrid PV/TE system, *Renewable Energy*, **87**, pp. 458–463.
18. Gao, H. B., Huang, G. H., Li, H. J., Qu, Z. G., Zhang, Y. J. (2016). Development of stove-powered thermoelectric generators: a review, *Appl. Thermal Eng.*, **96**, pp. 297–310.

19. Xie, Y., Wu, S. J., Yang, C. J. (2016). Generation of electricity from deep-sea hydrothermal vents with a thermoelectric converter, *Appl. Energy*, **164**, pp. 620–627.

20. Qian, B., Ren, F. (2016). Cooling performance of transverse thermoelectric devices, *Int. J. Heat Mass Transfer*, **95**, pp. 787–794.

21. Dresselhaus, M. S., Chen, G., Ren, Z., Dresselhaus, G., Henry, A., Fleurial, J. P. (2009). New composite thermoelectric materials for energy harvesting applications, *JOM*, **61**, pp. 86–90.

22. Dresselhaus, M. S., Chen, G., Tang, M. Y., Yang, R., Lee, H., Wang, D., Ren, Z., Fleurial, J. P., Gogna, P. (2007). New directions for low-dimensional thermoelectric materials, *Adv. Mater.*, **19**, pp. 1043–1053.

23. Casati, G., Mejia-Monasterio, C., Prosen, T. (2008). Increasing thermoelectric efficiency: a dynamical systems approach, *Phys. Rev. Lett.*, **101**, Article No. 016601.

24. Tritt, T. M., Bottner, H., Chen, L. (2008). Thermoelectrics: direct solar thermal energy conversion, *MRS Bull.*, **33**, pp. 366–368.

25. Boukai, A. I., Bunimovich, Y., Tahir-Kheli, J., Yu, J. K., Goddard, W. A., Heath, J. R. (2008). Silicon nanowires as efficient thermoelectrics materials, *Nature*, **451**, pp. 168–171.

26. Chowdhury, I., Prasher, R., Lofgreen, K., Chrysler, G., Narasimhan, S., Mahajan, R., Koester, D., Alley, R., Venkatasubramanian, R. (2009). On-chip cooling by superlattice-based thin-film thermoelectrics, *Nat. Nanotechnol.*, **4**, pp. 235–238.

27. Lee, J. H., Galli, G. A., Grossman, J. C. (2008). Nanoporous Si as an efficient thermoelectric materials, *Nano Lett.*, **8**, pp. 3750–3754.

28. Rusu, M., Caplanus, I., Mardare, D., Rusu, G. I. (2005). Electrical and thermoelectrical properties of some new conjugated polymers in thin films, *J. Optoelectgron. Adv. Mater.*, **7**, pp. 3149–3154.

29. Yoshino, H., Papavassiliou, G. C., Murata, K. (2008). Low-dimensional organic conductors as thermoelectric materials, *J. Therm. Anal. Calorim.*, **92**, pp. 457–460.

30. Jeng, M.-S., Yang, R., Song, D., Chen, G. (2008). Modeling the thermal conductivity and phonon transport in nanoparticle composites using Monte Carlo simulation, *J. Heat Transfer*, **130**, Article No. 042410.

31. Huang, X., Huai, X., Liang, S., Wang, X. (2009). Thermal transport in Si/Ge nanocomposites, *J. Phys. D: Appl. Phys.*, **42**, Article No. 095416.

32. Cox, C. A., Toberer, E. S., Levchenko, A. A., Brown, S. R. (2009). Structure, heat capacity, and high-temperature thermal properties of $Yb_{14}Mn_{1-x}Al_xSb_{11}$, *Chem. Mater.*, **21**, pp. 1354–1360.

33. Koza, M. M., Johnson, M. R., Viennois, R. (2008). Breakdown of phonon-glass paradigm in La-and Ce-filled Fe_4Sb_{12} skutterudites, *Nat. Mater.*, **7**, pp. 805–810.

34. Kim, H. J., Bozin, E. S., Haile, S. M. (2007). Nanoscale alpha-structural domains in the phonon-glass thermoelectric material beta-Zn_4Sb_3, *Phys. Rev. B*, **75**, Article No. 134103.

35. Kim, W., Wang, R., Majumdar, A. (2007). Nanostructuring expands thermal limits, *Nano Today*, **2**, pp. 40–47.

36. He, Z. M., Stiewe, C., Platzek, D., Karpinski, G., Mueller, E. (2007). Nano $ZrO_2/CoSb_3$ composites with improved thermoelectric figure of merit, *Nanotechnology*, **18**, Article No. 235602.

37. Heremans, J. P. (2005). Low-dimensional thermoelectricity, *Acta Phys. Pol. A*, **108**, pp. 609–634.

38. Baxter, J., Bian, Z. X., Chen, G., Danielson, D. (2009). Nanoscale design to enable the revolution in renewable energy, *Energy Environ. Sci.*, **2**, pp. 559–588.

39. Snyder, G. J., Toberer, E. S. (2008). Complex thermoelectric materials, *Nat. Mater.*, **7**, pp. 105–114.

40. Hochbaum, A. I., Chen, R. K., Delgado, R. D., Liang, W. J. (2008). Enhanced thermoelectric performance of rough silicon nanowires, *Nature*, **451**, pp. 163–165.

41. Biswas, K. G., Sands, T. D., Cola, B. A., Xu, X. F. (2009). Thermal conductivity of bismuth telluride nanowire array-epoxy composites, *Appl. Phys. Lett.*, **94**, Article No. 223116.

42. Guo, H., Screekumar, T. V., Liu, T., Minus, M., Kumar, S. (2005). Structure and properties of polyacrylonitrile/single wall carbon nanotube composite film, *Polymer*, **46**, pp. 3001–3005.

43. Sugar, J. D., Medlin, D. L. (2009). Precipitation of Ag_2Te in the thermoelectric material $AgSbTe_2$, *J. Alloys Compd.*, **478**, pp. 75–82.

44. Shinohara, Y., Isoda, Y., Imai, Y., Hiraishi, K., Oikawa, H. (2007). The effect of carrier conduction between main chains on thermoelectric properties of polythiophene, *Proc. 26th Int. Conf. Thermoelectrics*, **1**, pp. 405–407.

45. Yan, H., Ishida, T., Toshima, N. (2001). Thermoelectric properties of electrically conductive polypyrrole film, *Proc. 20th Int. Conf. Thermoelectrics*, **1**, pp. 310–313.

46. Hiroshige, Y., Ookawa, M., Toshima, N. (2007). Thermoelectric figure-of-merit of iodine-doped copolymer of phenylenevinylene with dialkoxyphenylenevinylene, *Synth. Met.*, **157**, pp. 467–474.

47. Gurrero, V. H., Wang, S., Wen, S., Chung, D. D. L. (2002). Thermoelectric property tailoring by composite engineering, *J. Mater. Sci.*, **37**, pp. 4127–4136.

48. Yin, X. H., Kobayashi, K., Yoshino, K., Yamamoto, H., Watanuki, T., Isa, I. (1995). Percolation conduction in polymer composites containing polypyrrole coated insulating polymer fiber and conducting polymer, *Synth. Met.*, **69**, pp. 367–368.

49. Wakim, S., Aich, B. R., Tao, Y., Leclerc, M. M. (2008). Charge transport, photovoltaic and thermoelectric properties of poly(2,7-carbazole) and poly(indolo[3,2-b]carbazole) derivatives, *Polym. Rev.*, **48**, pp. 432–462.

50. Eda, G., Chhowalla, M. (2009). Graphene-based composite thin films for electronics, *Nano Lett.*, **9**, pp. 814–818.

51. Hadziioannou, G., Malliaras, G. G. (2007). *Semiconducting Polymers*, 2nd ed., New York: Wiley-VCH, pp. 26–39.

52. Pinter, E., Fekete, Z. A., Berkesi, O., Makra, P., Patzko, A. (2007). Characterization of poly(3-octylthiophene)/silver nanocomposites prepared by solution doping, *J. Phys. Chem. C*, **11**, pp. 11872–11878.

53. Han, Y. Q., Lu, Y. (2009). Characterization and electrical properties of conductive/colloidal graphite oxide nanocomposites, *Compos. Sci. Technol.*, **69**, pp. 1231–1237.

54. Worsley, M. A., Kucheyev, S. O., Kuntz, J. D., Hamza, A. V., Satcher, J. H. (2009). Stiff and electrically conductive composites of carbon nanotube aerogels and polymers, *J. Mater. Chem.*, **19**, pp. 3370–3372.

55. Zhang, W., Li, J., Zou, L., Zhang, B., Qin, J. G. (2008). Semiconductive polymers containing dithieno[3,2-b:2′, 3′-d]pyrrole for organic thin film transistors, *Macromolecules*, **41**, pp. 8953–8955.

56. Gao, L., Wang, S., Liu, R., Zhai, S., Zhang, H., Wang, J., Fu, G. (2016). The effect of Ni doping on the thermoelectric transport properties of CdO ceramics, *J. Alloys Compd.*, **662**, pp. 213–219.

57. Ba, Y., Wang, Y., Wan, C., Norimatsu, W., Kusunoki, M., Ba, D., Koumoto, K. (2016). Thermoelectric properties of Nb-doped $(Nd_{0.55}Li_{0.36})TiO_3$ bulk ceramics with superlattice structure, *J. Alloys Compd.*, **664**, pp. 487–491.

58. Duran, C., Yildiz, A., Dursun, S., Mackey, J., Sehirlioglu, A. (2016). Thermoelectric characteristics of textured $KSr_2Nb_5O_{15}$ ceramics, *Scr. Mater.*, **112**, pp. 114–117.

59. Schulz, T., Töpfer, J. (2016). Thermoelectric properties of $Ca_3Co_4O_9$ ceramics prepared by an alternative pressure-less sintering/annealing method, *J. Alloys Compd.*, **659**, pp. 122–126.

60. Abhari, A. S., Abdellahi, M., Bahmanpour, M. (2016). The effects of Sn-substitution on thermoelectric properties of $In_{4-x}Sn_xSe_3$ ceramic, *Ceram. Int.*, **42**, pp. 5593–5599.

61. Li, Y., Liu, J., Hou, Y., Zhang, Y., Zhou, Y., Su, W., Zhu, Y., Li, J., Wang, C. (2015). Thermal conductivity and thermoelectric performance of $Sr_xBa_{1-x}Nb_2O_6$ ceramics at high temperatures, *Scr. Mater.*, **109**, pp. 80–83.

62. Hsieh, C. C., Hwang, C. S., Kuo, C. H. (2016). Room temperature thermoelectric properties of p-type $CuMn_{1.1-x}E_xO_2$ (E=Mg, Ca, Sr, x=0–0.2) ceramics, *Ceram. Int.*, **42**, pp. 2960–2968.

63. Zeng, C., Liu, Y., Lan, J., Ren, G., Lin, Y., Li, M., Nan, C. (2015). Thermoelectric properties of $Sm_{1-x}La_xBaCuFeO_5$ ceramics, *Mater. Res. Bull.*, **69**, pp. 46–50.

64. Li, Y., Liu, J., Zhang, Y., Zhou, Y., Li, J., Su, W., Zhai, J., Wang, H., Wang, C. (2016). Enhancement of thermoelectric performance of $Sr_{0.7}Ba_{0.3}Nb_2O_{6-\delta}$ ceramics by lanthanum doping, *Ceram. Int.*, **42**, pp. 1128–1132.

65. Delorme, F., Bah, M., Schoenstein, F., Jean, F., Zouaoui-Jabli, M., Monot-Laffez, I., Giovannelli, F. (2016). Thermoelectric properties of oxygen deficient $(K_{0.5}Na_{0.5})NbO_3$ ceramics, *Mater. Lett.*, **162**, pp. 24–27.

66. Flahaut, D. Allouche, J., Sotelo, A., Rasekh, S., Torres, M. A., Madre, M. A., Diez, J. C. (2016). Role of Ag in textured-annealed $Bi_2Ca_2Co_{1.7}O_x$ thermoelectric ceramic, *Acta Mater.*, **102**, pp. 273–283.

67. Abdellahi, M., Ghayour, H., Bahmanpour, M. (2015). Effect of process parameters and synthesis method on the performance of thermoelectric ceramics: a novel simulation, *Ceram. Int.*, **41**, pp. 6991–6998.

68. Hsieh, C. C., Hwang, C. S., Kuo, C. H., Qi, X. D., Hsiao, C. L. (2015). Fabrication and thermoelectric properties of $CuMn_{1+x}O_2$ (x=0~0.2) ceramics, *Ceram. Int.*, **41**, pp. 12303–12309.

69. Abdellahi, M., Bahmanpour, M., Bahmanpour, M. (2015). Modeling Seebeck coefficient of $Ca_{3-x}M_xCo_4O_9$ (M=Sr, Pr, Ga, Ca, Ba, La, Ag) thermoelectric ceramics, *Ceram. Int.*, **41**, pp. 345–352.

70. Zhu, Y. H., Su, W. B., Liu, J., Zhou, Y. C., Li, J., Zhang, X., Du, Y., Wang, C. L. (2015). Effects of Dy and Yb co-doping on thermoelectric properties of $CaMnO_3$ ceramics, *Ceram. Int.*, **41**, pp. 1535–1539.

71. Gao, F., Yang, S., Li, J., Qin, M., Zhang, Y., Sun, H. (2015). Fabrication, dielectric, and thermoelectric properties of textured $SrTiO_3$ ceramics prepared by RTGG method, *Ceram. Int.*, **41**, pp. 127–135.

72. Gonçalves, A. P., Lopes, E. B., Monnier, J., Bourgon, J., Vaney, J. B., Piarristeguy, A., Pradel, A., Lenoir, B., Delaizir, G., Pereira, M. F. C., Alleno, E., Godart, C. (2016). Fast and scalable preparation of tetrahedrite for thermoelectrics via glass crystallization, *J. Alloys Compd.*, **664**, pp. 209–217.

73. Liou, Y. C., Tsai, W. C., Yu, J. Y., Tsai H. C. (2015). Effects of Ti addition on properties of $Sr_2Nb_2O_7$ thermoelectric ceramics, *Ceram. Int.*, **41**, pp. 7036–7041.

74. Butt, S., Ren, Y., Farooq, M. U., Zhan, B., Sagar, R. U. R., Lin, Y., Nan, C. W. (2014). Enhanced thermoelectric performance of heavy-metals (M: Ba, Pb) doped misfit-layered ceramics: $(Ca_{2-x}M_xCoO_3)_{0.62}$ (CoO_2), *Energy Convers. Manage.*, **83**, pp. 35–41.

75. Zhu, Y., Wang, C., Wang, H., Su, W., Liu, J., Li, J. (2014). Influence of Dy/Bi dual doping on thermoelectric performance of $CaMnO_3$ ceramics, *Mater. Chem. Phys.*, **144**, pp. 385–389.

76. Rubešová, K., Hlásek, T., Jakeš, V., Huber, Š., Hejtmánek, J., Sedmidubský, D. (2015). Effect of a powder compaction process on the thermoelectric properties of $Bi_2Sr_2Co_{1.8}O_x$ ceramics, *J. Eur. Ceram. Soc.*, **35**, pp. 525–531.

77. Zhang, D. B., Zhang, B. P., Ye, D. S., Liu, Y. C., Li, S. (2016). Enhanced Al/Ni co-doping and power factor in textured ZnO thermoelectric ceramics prepared by hydrothermal synthesis and spark plasma sintering, *J. Alloys Compd.*, **656**, pp. 784–792.

78. Wang, H., Sun, X., Yan, X., Huo, D., Li, X., Li, J. G., Ding, X. (2014). Fabrication and thermoelectric properties of highly textured $Ca_9Co_{12}O_{28}$ ceramic, *J. Alloys Compd.*, **582**, pp. 294–298.

79. Wang, S., Lü, Q., Li, L., Fu, G., Liu, F., Dai, S., Yu, W., Wang, J. (2013). High-temperature thermoelectric properties of $Cd_{1-x}Pr_xO$ ceramics, *Scr. Mater.*, **69**, pp. 533–536.

80. Harizanova, S. G., Zhecheva, E. N., Valchev, V. D., Khristov, M. G., Stoyanova, R. K. (2015). Improving the thermoelectric efficiency of Co based ceramics, *Mater. Today Proc.*, **2**, pp. 4256–4261.

81. Zhu, C., Gong, J., Li, Z., Zhou, H., Tang, G. (2015). Investigation of the thermoelectric properties of La_2CuO_4 ceramics with the addition of Ag, *Physica B*, **456**, pp. 26–30.

82. Li, F., Wei, T. R., Kang, F., Li, J. F. (2014). Thermal stability and oxidation resistance of BiCuSeO based thermoelectric ceramics, *J. Alloys Compd.*, **614**, pp. 394–400.

83. Van Bel, J., Tong, C., Gan, R. N., Eshaghof, M., Nsavu-Nzau, C. D. N., Gan, Y. X. (2016). Processing and Seebeck effect measurement of a bismuth based alloy, *Adv. Res.*, **6**, Article No. 23400.

84. Zhao, Y., Chen, M., Zhang, Y., Xu, T., Liu, W. (2005). A facile approach to formation of through-hole porous anodic aluminum oxide film, *Mater. Lett.*, **59**, pp. 40–43.

85. Kong, L. B. (2005). Synthesis of Y-junction carbon nanotubes within porous anodic aluminum oxide template, *Solid State Commun.*, **133**, pp. 527–529.

86. Qiao, J., Pelletier, J., Li, N., Yao, Y. (2016). Insight on viscoelasticiy of $Ti_{16.7}Zr_{16.7}Hf_{16.7}Cu_{16.7}Ni_{16.7}Be_{16.7}$ high entropy bulk metallic glass, *J. Iron Steel Res. Int.*, **23**, pp. 19–23.

87. Zhang, T., Lin, X., Yang, H., Liu, Y., Wang, Y., Qiao, J. (2016). Tribological properties of a dendrite-reinforced Ti-based metallic glass matrix composite under different conditions, *J. Iron Steel Res. Int.*, **23**, pp. 57–63.

88. Shen, T. D., Xin, S. W., Sun, B. R. (2016). Low power loss in $Fe_{65.5}Cr_4Mo_4Ga_4P_{12}B_{5.5}C_5$ bulk metallic glasses, *J. Alloys Compd.*, **658**, pp. 703–708.

89. Su, C., Chen, Y., Yu, P., Song, M., Chen, W., Guo, S. F. (2016). Linking the thermal characteristics and mechanical properties of Fe-based bulk metallic glasses, *J. Alloys Compd.*, **663**, pp. 867–871

90. Gu, Y., Zheng, Z., Niu, S., Ge, W., Wang, Y. (2013). The seawater corrosion resistance and mechanical properties of $Cu_{47.5}Zr_{47.5}Al_5$ bulk metallic glass and its composites, *J. Non-Cryst. Solids*, **380**, pp. 135–140.

91. Wang, B. P., Yu, B. Q., Fan, Q. B., Liang, J. Y., Wang, L., Xue, Y. F., Zhang, H. F., Fu, H. M. (2016). Anisotropic dynamic mechanical response of tungsten fiber/Zr-based bulk metallic glass composites, *Mater. Des.*, **93**, pp. 485–493.

92. Guo, S. F., Chan, K. C., Xie, S. H., Yu, P., Huang, Y. J., Zhang, H. J. (2013). Novel centimeter-sized Fe-based bulk metallic glass with high corrosion resistance in simulated acid rain and seawater, *J. Non-Cryst. Solids*, **369**, pp. 29–33.

93. Gargarella, P., Pauly, S., Samadi Khoshkhoo, M., Kiminami, C. S., Kühn, U., Eckert, J. (2016). Improving the glass-forming ability and plasticity of a TiCu-based bulk metallic glass composite by minor additions of Si, *J. Alloys Compd.*, **663**, pp. 531–539.

94. Wang, L., Chao, Y. (2012). Corrosion behavior of $Fe_{41}Co_7Cr_{15}Mo_{14}C_{15}B_6Y_2$ bulk metallic glass in NaCl solution, *Mater. Lett.*, **69**, pp. 76–78.

95. Lee, K. S., Kim, S., Lim, K. R., Hong, S. H., Kim, K. B., Na, Y. S. (2016). Crystallization, high temperature deformation behavior and solid-to-solid formability of a Ti-based bulk metallic glass within supercooled liquid region, *J. Alloys Compd.*, **663**, pp. 270–278.

96. Zhao, G. H., Aune, R. E., Mao, H., Espallargas, N. (2016). Degradation of Zr-based bulk metallic glasses used in load-bearing implants: a tribocorrosion appraisal, *J. Mech. Behav. Biomed. Mater.*, **60**, pp. 56–67.

97. Zhang, J., Estévez, D., Zhao, Y. Y., Huo, L., Chang, C., Wang, X., Li, R. W. (2016). Flexural strength and Weibull analysis of bulk metallic glasses, *J. Mater. Sci. Technol.*, **32**, pp. 129–133.

98. Wang, Y. B., Li, H. F., Zheng, Y. F., Li, M. (2012). Corrosion performances in simulated body fluids and cytotoxicity evaluation of Fe-based bulk metallic glasses, *Mater. Sci. Eng. C*, **32**, pp. 599–606.

99. Zameer Abbas, S., Ahmad Khalid, F., Zaigham, H. (2016). Indentation and deformation behavior of FeCo-based bulk metallic glass alloys, *Mater. Sci. Eng. A*, **654**, pp. 426–435.

100. Zhou, K., Liu, Y., Pang, S., Zhang, T. (2016). Formation and properties of centimeter-size Zr–Ti–Cu–Al–Y bulk metallic glasses as potential biomaterials, *J. Alloys Compd.*, **656**, pp. 389–394.

101. Hu, X., Qiao, J., Pelletier, J. M., Yao, Y. (2016). Evaluation of thermal stability and isochronal crystallization kinetics in the $Ti_{40}Zr_{25}Ni_8Cu_9Be_{18}$ bulk metallic glass, *J. Non-Cryst. Solids*, **432**, pp. 254–264.

102. Cui, J., Li, J., Wang, J., Kou, H. (2016). Deformation behaviors of a Ti-based bulk metallic glass composite in the dendrite softening region, *Mater. Sci. Eng. A*, **653**, pp. 1–7.

103. Nollmann, N., Binkowski, I., Schmidt, V., Rösner, H., Wilde, G. (2016). Impact of micro-alloying on the plasticity of Pd-based bulk metallic glasses, *Scr. Mater.*, **111**, pp. 119–122.

104. Shamlaye, K. F., Laws, K. J., Ferry, M. (2016). Supercooled liquid fusion of carbon fibre-bulk metallic glass composites with superplastic forming properties, *Scr. Mater.*, **111**, pp. 127–130.

105. Gostin, P. F., Oswald, S., Schultz, L., Gebert, A. (2012). Acid corrosion process of Fe-based bulk metallic glass, *Corros. Sci.*, **62**, pp. 112–121.

106. Cui, J., Li, J., Wang, J., Li, L., Kou, H. (2016). Deformation behavior of a Ti-based bulk metallic glass composite in the supercooled liquid region, *Mater. Des.*, **90**, pp. 595–600.

107. White, F. M. (2008). *Fluid Mechanics*, 8th ed., Boston: McGraw-Hill, p. 226.

108. Mattis, D.C. (1981). *The Theory of Magnetism I*, Berlin: Springer-Verlag, p. 16.

109. Lorrain, P., Corson, D. R., Lorrain, F. (2001). *Fundamentals of Electromagnetic Phenomena*, New York: W. H. Freeman and Company, p. 62.

110. Mattis, D. C. (2006). *The Theory of Magnetism Made Simple*, Hackensack: World Scientific, p. 22.

111. Majlis, N. (2000). *The Quantum Theory of Magnetism*, Hackensack: World Scientific, p. 189.

112. Jiles, D. (1991). *Introduction to Magnetism and Magnetic Materials*, New York: Chapman and Hall, p. 55.

113. Morrish, A. H. (2001). *The Physical Principles of Magnetism*, Piscataway: IEEE Press, p. 23.

114. Mott, N. F. (1987). *Conduction in Non-Crystalline Materials*, Oxford: Clarendon Press, p. 53.

115. Ince, E. L. (1956). *Ordinary Differential Equations*, New York: Dover Publications, pp. 23–25.

Chapter 10

Nanocasting Thermoelectric Composite Materials

This chapter is on manufacturing multicomponent organic/inorganic composite materials containing nanotubes, nanofibers, and nanoparticles by nanocasting for energy conversion. The effect of combined electric, magnetic, and/or mechanical forces on the distribution of nanoscale particles, tubes, or fibers in matrix materials is discussed. The manufactured composite materials are tested for thermoelectric energy conversion. The nanocasting manufacturing process during which electromagnetic and mechanical actions coexist is introduced first. Then characterization of the properties of the thermoelectric materials with multiple components and complex structures is performed. Semiconducting polymer (polyaniline, PANI)–based multilayer composite nanofibers or nanorods in carbon-doped titania nanotubes are made by the nanocasting manufacturing technique. The titania nanotube template serves as the casting mold for electromagnetic force–assisted centrifugal nanocasting. Self-organized titania nanotubes are made by electrochemical oxidation of pure titanium in a solution containing fluorine ions. High-temperature heat treatment is conducted to dope titania nanotubes with carbon. Also introduced is a multilayered cobalt oxide nanotube–loaded titania coaxial nanorod composite made by nanocasting. The Seebeck coefficient of the nanorod composite materials is measured.

Nanomaterials for Thermoelectric Devices
Yong X. Gan

ISBN 978-981-4774-98-7 (Hardcover), 978-0-429-48872-6 (eBook)
www.panstanford.com

10.1 Introduction

Nanocasting refers to the new manufacturing technology for generating well-ordered nanostructures containing nanofibers, nanowires (NWs), nanorods, coaxial nanotubes (NTs), and pores. It is different from conventional casting. In the conventional casting process, the solidification of a molten metal occurs in a die [1]. It is widely used as a cost-effective approach to making machine parts [2]. Casting may be integrated into other manufacturing processes, for example, hot rolling and extrusion, which is called semisolid casting. In this combined process, a molten metallic alloy, such as a silicon containing aluminum alloy, is poured into a mold to form an ingot. The ingot is under the pulling and subjected to a hot-rolling process. To increase the cooling rate of solidification, a molten alloy may be injected into a water-cooled mold to be quickly cooled down to a mixed state having both liquid and solid components. In the subsequent process, the semisolid slurry is subjected to extrusion or rolling before it changes completely into a solid. In such a combined manufacturing process, the dendritic growth can be suppressed. Consequently, the chemical composition of the cast alloy is more uniform and the mechanical properties of the part can be improved [3].

To make nanoscale architectures and components, nanocasting technology has been developed. Nanocasting has found major applications in manipulation of nonmetallic materials such as polymers and ceramics. For example, nanoporous polymer structures were made by casting oligomeric surfactants and block copolymers as reported [4–8]. In literature [9], a pseudopolyrotaxanes template synthesized from α-cyclodextrin (CD) and polyamines was used to nanocast porous silica structures. Micro- and nanoscale hierarchical two-dimensionally ordered porous array structures can be made via nanocasting by the use of a monolayer colloidal template. The polystyrene (PS) bead with a diameter of 1 micron was found to form two-dimensionally ordered arrays in a proper solution [10]. These ordered arrays were demonstrated as nanocasting templates or molds to make Co_3O_4 hierarchical structures.

Porous structures containing nanocrystalline materials can be obtained by cocasting gels and polymer microspheres as shown in Ref. [11]. In Ref. [12], how polymer colloidal crystal beads self-assembled into 3D templates for nanocasting mold application is presented. The 3D template structures consisting of microspheres may be assembled on different substrates, including glass slides, quartz plates, and other ceramics, by the vertical deposition approach [13]. Polymethyl methacrylate (PMMA) colloidal crystal monoliths were used as the starting templates for nanocasting [14, 15]. Zhang and Wong [16] used a track-etched polycarbonate (PC) polymer membrane as the template for casting 1D nanostructures. Various discrete motifs and arrays of crystalline and pure semiconducting transition metal sulfides, including CuS, PbS, and CdS NWs, were made via template-directed nanocasting. The process consists of nanocasting octadecyltetrachlorosilane self-assembled monolayers (OTS-SAMs) on the front and back surfaces of the PC template. OTS-SAMs serve as passivation layers [17].

An anodic aluminum oxide (AAO) template was found to be prepared fairly easily [18–31], and it was used to cast nanofibers or NTs. Even some plant leaves can be used as the casting templates or molds. Lotus and colocasia leaves are two examples. In Refs. [32, 33], thin polymeric films were cast onto lotus and colocasia leaves to make a replica of the surface structures. Many ordered structures are made from mesoporous silica. Such ordered structures have been extensively used for nanocasting [34–66]. Since the pore size and shape of mesoporous silica materials can be controlled, the morphological features of the cast porous structures using such silica molds could be very different. Nanocasting porous transition metal oxides, such as Cr_2O_3, Co_3O_4, In_2O_3, NiO, CeO_2, WO_3, Fe_2O_3, and MnO_2, using these mesoporous silica templates was summarized by Yue and Zhou [67]. The pore size can be as fine as less than 10 nm. Some of the porous transition metal oxide nanomaterials have found applications as catalysts, Li ion rechargeable battery components, and gas sensors [68].

Many researchers have studied porous carbon as an intermediate template material for nanocasting [69–87]. To make a porous carbon template, the manufacturing process involves the preparation of a micromesoporous carbon mold using an aggregated silica sphere–assembled template and a carbon source. The diameter of the silica spheres is typical at the 10 nm level [69]. Lu et al. [70] prepared ordered mesoporous carbon containing highly dispersed copper sulfur compounds in the carbon framework via nanocasting. The mesoporous silica, SBA-15, was used as the starting template, and copper (II) phthalocyanine-tetrasulfonic acid tetrasodium salt (PcS) was used as the precursor. At the pyrolysis temperatures lower than 600°C, stable nanocast carbon was obtained. If the treatment temperature is higher than 600°C, highly dispersed copper sulfur compounds are formed in the carbon framework. Porous carbon was also used as the template for nanocasting oxides. Lai et al. [71] synthesized crystalline mesoporous indium oxide using a mesoporous carbon (CMK-3) as the hard template.

Wu et al. [72] described the formation of unimodal and bimodal mesoporous carbon via nanocasting. Methylated β-CD and tetramethyl orthosilicate (TMOS) were used as the starting materials. By controlling the ratio of CD to TMOS, both unimodal and bimodal porous carbons can be obtained. The mechanism for both types of porous carbon may be explained as follows: With a proper CD/TMOS ratio, the silica frame forms a mesoporous structure. By increasing the TMOS content, many silica gel frameworks in CD/silica wet gels could collapse due to lack of support for CD during the drying process and then generate many large silica nanoparticles. Consequently, the silica component in the dried CD/silica composites with low CD/TMOS ratios has two types of structures. One is the gel skeletal structure, and the other is the nanoparticle structure. A simpler one-step nanocasting method to prepare bimodal mesoporous carbon was also shown in Ref. [72]. Carbon templates can be made in both fiber and tubular forms [88]. A freestanding carbon monolith was reused for the second casting step with a porous alumina template. By casting polymers into the pores, alumina/carbon nanotube (CNT) composites can be made. Popp et al. [88] suggested that prior to the removal of the alumina using hydrofluoric acid (HF), the inner CNT space should be protected by an acrylate polymer. After the second casting step, and thermal removal of the acrylate layer, the CNT-reinforced material took the new composite structure. This structure duplicates the original alumina template. Carbon microtubes were obtained by casting [89]. In Ref. [90], the functionalization of mesoporous materials, including porous carbon, was introduced. The incorporation of NiO nanocrystals with different contents on the inner wall of the ordered porous carbon (CMK-3) was carried out in the synthesis procedure of combining the nanocasting strategy with the thermolysis as described by Li et al. [91]. Hollow spherical carbon with a mesoporous shell (HCMSC) was obtained by nanocasting [92].

Zeolites are minerals with regularly aligned porous arrays. The composition of zeolites could change in a large range, and their structure could be very complicated. Zeolites are mainly used in heterogeneous catalysis. They have also been considered as nanocasting templates for making various functional materials and structures [93–96]. Srinivasu et al. [93] made a microporous carbon nitride (CN) material with very high surface area and a large fraction of pore volume by nanocasting technique using the MCM-22 zeolite as a template through a polymerization reaction between ethylenediamine and carbon tetrachloride. Parmentier et al. [94] investigated the effect of zeolite crystal size on the structure and properties of carbon replicas made by nanocasting. Zeolite particles with different sizes (from 100 nm to 1 μm) were used to form the nanocasting molds for manufacturing microporous carbon. The procedures include high-temperature infiltration and calcination of a carbon source precursor and etching of the zeolite templates. Propylene was used as the carbon source, and the pyrolysis was carried out at 800°C for 2 h. The carbon replica was obtained by dissolution of the zeolite with 40% HF. In the work performed by Ducrot-Boisgontier et al. [95], different precursors were used to make a carbon replica on the zeolite (EMC-2) mold. Acetylene, furfuryl alcohol, and acetonitrile were used as the carbon sources. The nanocasting consisted of a single or double infiltration procedure. The pore size ranged from 1.0 to 1.5 nm.

Commercially available PS beads were used as the nanocasting molds [97]. Shang et al. [98] used polyethylene glycol (PEG) as the scaffold for nanocasting oxides. A sonochemical approach was used to synthesize $BiVO_4$ photosensitive catalyst with high photocatalytic activity. Xing and Rankin [99] used mixed surfactant systems to control the pore structures of mesoporous metal oxides. A glucopyranoside surfactant with a cationic surfactant that readily forms liquid crystalline mesophases was used in their work. A large area of hexagonal phase, as well as cubic, lamellar, and solid surfactant phases, was found. The ternary phase diagram was used to predict the synthesis of thick mesoporous silica films via a direct liquid crystal casting technique.

Nanocasting can be combined with other micro- and nanomanufacturing techniques, for example, lithography and nanoimprinting, to form fine structures [100] or to do fine pattern transfer [101]. Sogo et al. [100] proposed using fluorinated polymer and nanocasting lithography (NCL) to reproduce fine patterns. The reproduced antireflection structures were examined by a scanning electron microscope. A typical scanning electron microscopic (SEM) image shows that the fine feature has the size of 250 nm. NCL affords you the advantage of using the conventional contact printing tool to make nanoscale features. The processing can be carried out under ambient conditions. Thus, the NCL is easy to be implemented. Hirai et al. [101] applied this process for transferring fine patterns using various polymers such as PC, PMMA, and polylactic acid (PLA). The substrates used were quartz PLA and PC plates. High aspect ratio patterns were obtained on these substrates.

Nanocasting materials have found wide applications [102–121]. Catalysis is one of the most important applications [102–108]. For example, mesoporous silica was used as the starting template for porous carbon deposition from sucrose. A porous carbon mold was used to nanocast titanium silicalite-1 (TS-1) for the catalytic oxidation of aromatic thiophene [102]. Rao et al. [103] nanocast Cs-based nanocrystal salts using a silica nanoporous mold (SBA-15). The CsHPW nanocasts showed enhanced catalytic activity in the isopropanol dehydration reactions.

Nanocasting was used to build energy converters. One of the examples is the biophotofuel cell electrode. Although the photobiological process for hydrogen production has been studied for a long time [121–127], there are many problems to be solved. For example, the efficiency of energy conversion is low because the performance of the electrode needs to be improved. It has been shown that both quantum efficiency and hydrogen generation rate of biophotofuel cells can be increased by using new materials, such as semiconducting oxides [128–131]. The advantage of using nanomaterials [132, 133], that is nanoparticles, is that nanoparticles have much higher surface areas than bulk materials. Nevertheless, the agglomeration of particles is a challenge that needs to be resolved to ensure a high surface area of the electrode. Two ways have been proposed to solve the agglomeration problem on the basis of the studies of thermoelectric (TE) energy conversion nanomaterials [134, 135]. The first way is to cause nanoscale phases to grow out of the plane, along certain directions, to form nanoscale fractals or dendrites, which prevents the agglomeration of nanoparticles on substrates. The other method is to use nanoporous substrates or templates because they can confine the growth of nanocrystals to either inside the wall of the nanopores or to the surface of the substrates to prevent the agglomeration of nanoparticles. Using a nanoporous membrane as the template for nanocasting cobalt oxide and sulfide was performed, and the growth of the cobalt oxide followed the controlled way. It must be pointed out that cobalt compounds have attracted great interest because of the promising applications in catalysis, lithium batteries (as the cathode material), H_2S gas removal devices, photothermal, and photovoltaic energy conversion systems [136–156]. Nanocasting cobalt particles into oxide porous templates for making composite materials is presented in Ref. [157].

The centrifugal nanocasting or spinning concept has been studied for many years [158–162]. This technique has been used for making polymer nanofibers [158, 160, 162]. In principle, it can be applied for manufacturing ceramic-based fibers [159, 161]. By extending the centrifugal casting process concept to inorganic systems, it has become possible to synthesize fibers as suitable organic-inorganic composites. Up to now, there is limited research work done on both electric and magnetic force–assisted centrifugal nanocasting. The challenges of this new technology include the following: how to use a single-step process to manufacture large-scale fiber-reinforced materials; how to increase the production rate of the nanostructures while keeping the cost associated with the manufacturing process reasonable; and most importantly, how to manufacture self-supported composites with controlled architectures for use in functional devices such as TE energy conversion units.

The main objective of this chapter is to introduce the latest research results of the electromagnetic force–assisted centrifugal casting manufacturing process. How to use this new approach to make organic-inorganic composites is presented. Nanocasting as a potentially viable, scalable manufacturing process ensuring high process yield, process and product repeatability and reproducibility, and optimized quality control for producing functional nanocomposites for high-efficiency TE energy conversion is discussed.

10.2 Materials and Experimental

The experimental section can be divided into several parts. An oxide NT, that is a TiO_2 NT, was processed first. Next, the electromagnetic force–assisted centrifugal casting system was made based on a precision lapping/polishing machine. Then, TE composites were made using the centrifugal casting machine. The Seebeck coefficient of the composites was measured using an HP 34401A multimeter. The flowchart of the work is given in Fig. 10.1.

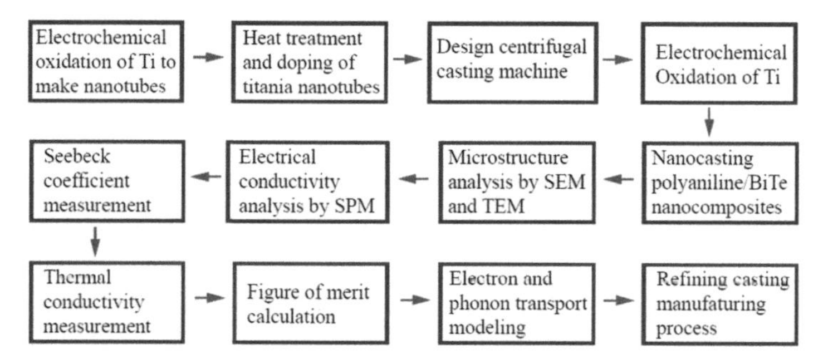

Figure 10.1 Flowchart showing nanocasting and characterization activities.

10.2.1 TiO_2 Nanotube Processing

TiO_2 NTs were generated on pure titanium by anodic oxidation. The titanium (Ti) thin sheet specimens 50 × 5 × 0.1 mm in size were cut from an as-supplied Ti foil with a thickness of 0.1 mm. Self-organized TiO_2 NTs were prepared by electrochemical oxidation of the Ti samples in an electrolyte with 1.5% NH_4F + 90% glycerol + 8.5% water in weight. A two-electrode cell was used for the electrochemical oxidation of Ti at room temperature (25°C). The anode and the cathode were both Ti with the same dimensions. The distance between the two electrodes was 20 mm. The operation voltage was 50.0 V, and the electrochemical oxidation time was 6 h. After electrochemical oxidation, the samples were rinsed in deionized water and air-dried. The surface of the anode was completely covered by TiO_2 NTs, as revealed by electron microscopic analysis. After the TiO_2 NTs were obtained, high-temperature annealing (at 500°C) for 2 h was performed to crystallize the TiO_2 and to dope the NTs with carbon. Then, the specimens were cooled down naturally and used as the mold for centrifugal nanocasting. Another way to process TiO_2 NTs is by the chemical reaction approach. Titanium oxide NTs were formed on the inner wall of the pores in the AAO template, as will be discussed later in this chapter.

10.2.2 Electromagnetic Force–Assisted Centrifugal Nanocasting

Figure 10.2a shows the concept of the electromagnetic force–assisted centrifugal nanocasting system. A precision auto lapping/polishing machine was used as the main body. This machine contains a rotating platform whose speed can be well controlled. A Shimpo tachometer was used to measure its rotating speed. The TiO_2 NT specimens were put at the two ends of a plastic pipe being fixed on the rotating platform. A DC voltage of 10 kV was applied across the aniline solution. Aniline was electrochemically polymerized in the TiO_2 NTs because the titanium plates, as anodes, were connected by two carbon fiber brushes to the outer ring, which served as the positive electrode of the DC power source. The Nd-Fe-B rare earth magnets on two sides generated a steady magnetic field in space. Under the combined electric and centrifugal mechanical forces, the polymerized PANI was cast into nanofibers within the TiO_2 NTs. The centrifugal casting principle is shown in Fig. 10.2b.

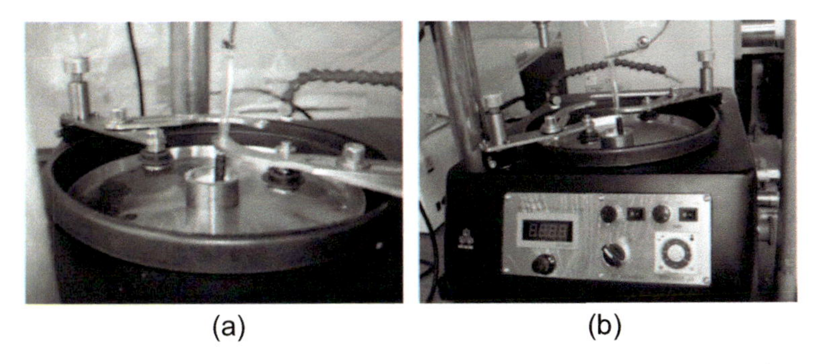

(a) (b)

Figure 10.2 Electromagnetic force–assisted centrifugal nanocasting machine: (a) the rotating platform holding the electrospinning unit and (b) casting under combined forces.

10.2.3 Nanocasting Using an Anodic Aluminum Oxide Template

Nanocasting ceramic nanocomposites is described in this part. In brief, TiO_2 NTs were generated through chemical reactions at the inner wall of the nanoporous AAO template. The pore diameter of AAO is about 20 nm. Before the synthesis of the TiO_2 NTs following the reaction below (Eq. 10.1), a self-assembled monolayer was deposited on both sides of the AAO membrane to inhibit the deposition of the TiO_2 particles at the top and bottom surfaces of the AAO. Consequently, TiO_2 formed NTs around the inner wall of the pores rather than particle aggregations occurring on the two side surfaces of the AAO. The self-assembled monolayer was prepared by mixing 10 mM octadecyltetrachlorosilane and 50 mL hexane solution. After that, TiO_2 NTs were generated by immersing the AAO membrane with the monolayer into the mixed 10 mL aqueous solution containing 0.05 M $(NH_4)_2TiF_6$ and 0.1 M H_3BO_3 for about 30 min. to guarantee TiO_2 NTs of a suitable thickness within the nanopores. The function of the H_3BO_3 is to accelerate the TiO_2 deposition through the reaction described by the following equations. Then the deposited AAO template was washed by deionized (DI) water several times and air-dried. The TiO_2 NT–inserted AAO was then heated at 550°C for 2 h when the TiO_2 NT arrays were finally obtained. The hydrolysis reaction associated with the initiation and growth of TiO_2 NTs is expressed by the Eq. 10.1:

$$[TiF_6]^{2-} + 2H_2O = TiO_2 + 6F^- + 4H^+ \quad (10.1)$$

To shift the hydrolysis to the right side of the equation, the F^- ions should be consumed to ensure the formation of the TiO_2. This can be achieved by adding boric acid through Eq. 10.2:

$$H_3BO_3 + 4H^+ + 4F^- = [BF_4]^- + H_3O^+ + 2H_2O \quad (10.2)$$

To enhance the TE performance of the composite material, Ag nanoparticles were deposited in situ into the TiO_2 NTs to increase the electrical conductivity. To obtain fine Ag nanoparticles at the surface of the TiO_2 NTs, TiO_2 AAO was dipped into a 0.05 M 10 mL $AgNO_3$ solution. After being washed by DI water and dried, it was heated at 500°C for 1 h. Because $AgNO_3$ is quite easy to decompose, Ag precipitations were generated and uniformly dispersed onto the inner wall of the TiO_2 NTs. The decomposition of $AgNO_3$ follows Eq. 10.3:

$$2AgNO_3 = 2Ag + 2NO_2\uparrow + O_2\uparrow \quad (10.3)$$

A self-assembled monolayer was still used to prevent the uneven deposition. The self-assembled monolayer was the same as mentioned before. The AAO template with TiO_2 NTs was immersed into the 0.05 M 10 mL Co $(NO_3)_2$ solutions for 20 min. to obtain the TiO_2-CoO nanocables. After the Co^{2+} ions and NO_{3-} ions completely permeated the TiO_2 NTs, the template was washed by DI water and dried. Subsequently, it was heated at 450°C for 30 min. to obtain the CoO-TiO_2 bilayer NTs. These tubes have the core-shell nanocable structure with a fairly smooth surface. $Co(NO_3)_2$ decomposes to CoO following the reaction of Eq. 10.4:

$$Co\ (NO_3)_2 = 2CoO + 4NO_2\uparrow + O_2\uparrow \quad (10.4)$$

It is noticed that by dissolving the AAO template into the solution with H^+ and F^- ions, a TiO_2-CoO nanocomposite can be obtained. The dissolution of the AAO substrate in the solution follows the reaction as described by Eq. 10.5:

$$Al_2O_3 + 12H^+ + 12F^- = 2H_3AlF_6 + 3H_2O \quad (10.5)$$

Morphology studies [157] show that TiO_2 nanostructure deposition causes AAO nanopore expansion. The original average pore size of the AAO template is 20 nm; however, after the growth of the TiO_2 at the inner wall of the pores, the pore size is increased to about 200 nm. This may be caused by the eroding function of H_3BO_3 and HF. Under a scanning electron microscope, it can be seen clearly that some thin walls of the NTs have been corroded and have fused with thicker adjacent NTs to form expanded NTs [157]. It is reported that both NaOH and H_3PO_4 can also expand the size of the NTs and NaOH has a faster etching speed. In our case, on the other hand, either HF or H_3BO_3 has a pore expanding effect similar to NaOH or H_3PO_4. The mechanism of expansion of NTs can be explained by the reactions illustrated by Eqs. 10.6a and 10.6b:

$$Al_2O_3 + 12OH^- + 12Na^+ = 2Na_3AlO_3 + 3H_2O \quad (10.6a)$$

$$Al_2O_3 + 12H^+ + 4PO_4^{3-} = 2Al(H_2PO)_4 + 3H_2O \quad (10.6b)$$

Using the energy-dispersive X-ray spectroscopy (EDS), Su et al. analyzed the compositions of the nanocomposite containing TiO_2

NTs [157]. The results show that the nanocast composite material contains major elements from TiO_2 NTs, Ag particles, and the AAO template. Under a transmission electron microscope, a darker region was observed, representing the deposited CoO core with the diameter of about 10 nm. Smaller black particles scattered within the NTs were found to be Ag particles. The number of Ag particles and the thickness of the TiO_2-CoO nanocables can be adjusted by the liquid-phase deposition time in the corresponding solutions. That is to say, if we repeat the deposition process, multilayered nanocables can be obtained with potential tailored structures and TE performances. By analyzing the electron diffraction pattern of the TiO_2-CoO nanocables, it has been proven that the CoO core has a crystalline structure.

10.3 Structure of Nanocast Composite Materials

A preliminary study showed that without combined electric and centrifugal forces, electrochemically polymerized PANI starts growing just at the surface of the titanium dioxide NTs, as revealed by both the SEM image of Fig. 10.3a and the transmission electron microscopic (TEM) image of Fig. 10.3b. Because the aniline solution is hydrophobic, it tends to stay on the top surface of the titania NTs, which are hydrophic. Therefore, electrocentrifugal nanocasting has to be used to draw the aniline into the NTs under coupled electric and mechanical forces.

TE Bi-Te metallic nanoparticles were made by the Galvanic displacement method similar to that reported in Refs. [163, 164]. In Refs. [163, 164], a Ni-Fe alloy was used as the starting material for Gavanic reactions. In this work, Ni and Co nanoparticles were used as the precursors to form core-shell particles due to the better controllability of the Galvanic reactions. After the Galvanic reactions, the surface layer of the particles contains a Bi-Te alloy and the core keeps the original composition of either Ni or Co. The Bi-Te alloy was used because of its strong Seebeck effect in a relatively wide temperature range.

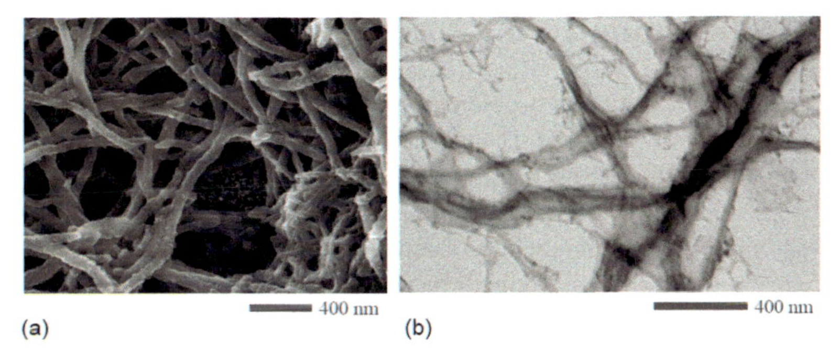

Figure 10.3 Images of polyaniline nanofibers on titanium dioxide nanotubes: (a) under SEM, and (b) under TEM. Republished with permission of ASME, from Ref. [165] (2015); permission conveyed through Copyright Clearance Center, Inc.

Since Ni or Co tends to be attracted by magnetic forces, the Bi-Te/Ni or Bi-Te/Co shell-core nanoparticle clusters (as shown in Fig. 10.4a) in the aniline solution held by the plastic container (as shown in Fig. 10.4b) move outwardly toward the TiO_2 NTs. It must be pointed out that the size of these nanoparticles, 4 nm, as seen from Fig. 10.4a, is much smaller than the inner diameter of the NTs, which is about 120 nm, as can be seen from Fig. 10.4c. Therefore, in the magnetic field, centrifugal casting allows the TE nanoparticles to go into the titanium NTs.

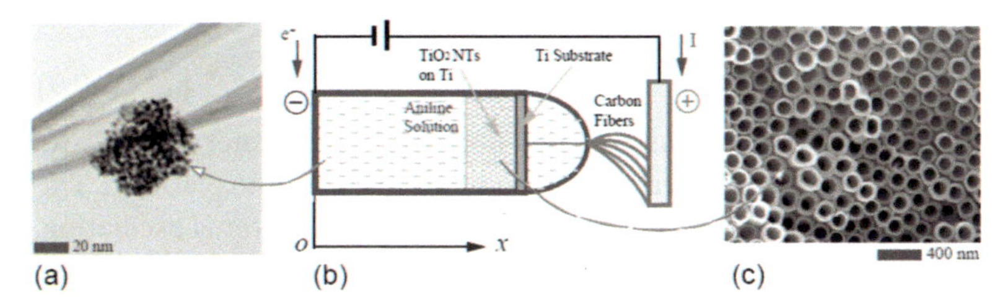

Figure 10.4 (a) TEM image of a Bi-Te/Ni shell-core nanoparticle cluster generated by Galvanic displacement, (b) Bi-Te/Ni in am aniline solution, and (c) SEM image of titanium dioxide nanotubes. Republished with permission of ASME, from Ref. [165] (2015); permission conveyed through Copyright Clearance Center, Inc.

10.4 Manufacturing Process Analysis

One of the fundamental problems associated with the manufacturing process is how fast the TE nanoparticles move into the titanium dioxide NTs under the combined electromagnetic and mechanical forces. Referring to the coordinates illustrated in Fig. 10.5a, the position of the nanoparticle cluster is designated as (x, y); the velocity components along different directions are $\dot{x}$ and $\dot{y}$, and the corresponding acceleration components are $\ddot{x}$ and $\ddot{y}$. Since the particle is initially negatively charged from the feeding slot, an electric force F_e exists. This force is proportional to both the electric charge carried by the nanoparticle cluster and the intensity of the electric field between the positive and negative electrodes generated by the power source. F_e can be expressed by Eq. 10.7.

$$F_e = \frac{qV}{l-x}, \tag{10.7}$$

where q is the charge; V is the voltage of the power source; and l is the distance between the positive and negative electrodes, which can be approximated by the radius of the rotating platform. Due to the centrifugal action, the mechanical force is given by:

$$F_c = \omega x^2, \tag{10.8}$$

where ω is the angular velocity of the platform. The magnetic force in the transverse direction (Lorentz force) as a function of the speed $\dot{x}$ and time t may be expressed as

$$F_m = q\dot{x}B\cos(2\omega t), \tag{10.9}$$

where B is the intensity of the magnetic field generated by the magnets. The cosine term is due to the rotation of the platform. The action of the magnetic field on the nanoparticle cluster is oscillating. The factor 2 in the cosine function is related to the two magnets on a circle. The viscous drag from the aniline in term of $v\dot{x}$ should also be considered, where v is the viscosity of the solution. The problem associated with the nonlinear dynamics in the manufacturing process can be formulated by the following system ordinary differential equations.

For the TE nanoparticle

$$\ddot{x} + v\dot{x} - \omega x^2 - \frac{qV}{l-x} = 0, \tag{10.10a}$$

and

$$\ddot{y} - q\dot{x}B\cos(2\omega t) = 0 . \qquad (10.10b)$$

For the aniline monomer and/or PANI polymer, similar approaches can be used.

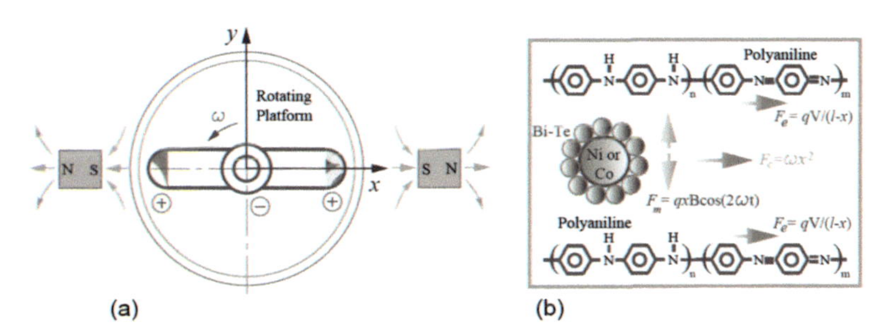

Figure 10.5 (a) The centrifugal nanocasting platform and the coordinates for analysis, and (b) the free body diagrams for different entities in the composite materials.

Once the differential equations are formulated and refined, the phase portraits of the nonlinear dynamic system can be generated. The phase curves may be plotted, and the related separatrix can be identified by the approaches presented in earlier work [166–168]. In addition, bifurcation properties and the corresponding bifurcation points can be determined. The analytical results can be used to examine the effect of the manufacturing parameters on the production rate and the uniformity of the composite materials made in the nanocasting process. For example, it is possible to find out how variations in the rotating speed of the platform, the intensity of the magnetic field, and the viscosity of the solution change the production rate of the nanocomposites and how the morphology and distribution of the nanoparticles within the composite materials are changed by these processing or manufacturing parameters. Although, our preliminary studies via trial-and-error experiments show that sometimes the cast materials could go into the NTs uniformly, as shown in Figs. 10.6a and 10.7a. Figures 10.7b and 10.7c verify the cast materials within the NTs and with the desired composition. However, at other times the cast materials cannot fill all the NTs. In other words, the repeatability of the casting process has to be improved. To achieve the complete filling state, the effect of the magnetic field needs to be examined. The oscillating magnetic forces acting on the nanoparticle cluster and the polymer generate vibration and stirring effects in small areas. It is expected to improve the filling process. The bifurcation conditions for the system, that is Eqs. 10.10a and 10.10b, were studied. The results provided some insight into the optimized filling conditions.

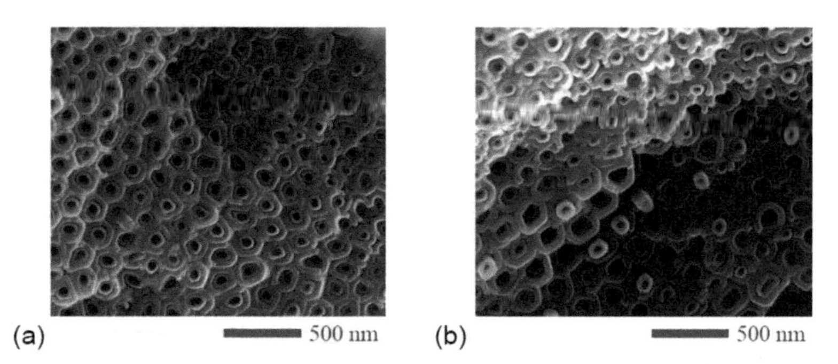

Figure 10.6 SEM images showing the filling state of cast materials in self-organized titanium dioxide nanotubes: (a) complete filling with higher uniformity and (b) partial filling with lower uniformity.

Figure 10.7 TEM morphological and elemental analysis of nanocast TiO_2 nanotubes: (a) cast materials in TiO_2 nanotubes to form a multilayered structure, (b) inner wall morphology at high magnification showing different phases, and (c) EDX diffraction spectrum confirming Ti and O elements from the nanotubes and Cu from the mesh for sample holding. Republished with permission of PERGAMON, from Ref. [157] (2010); permission conveyed through Copyright Clearance Center, Inc.

It must be pointed out that metallic and ceramic nanoparticles were also nanocast into TiO_2 NTs. In Fig. 10.8a, the TEM morphology of multilayered TiO_2/CoO nanofibers with incorporated nanoparticles prepared from nanocasting without a magnetic field is presented. As a control experiment, silver nanoparticles were used. After the nanofibers were broken, the Ag nanoparticles at the interface of

TiO_2/CoO were revealed, as shown in Fig. 10.8b. The average particle size is about 5 nm. The energy-dispersive X-ray (EDX) diffraction spectrum confirmed the major composition of Ti, Ag, Co, and O, as shown in Fig. 10.8c. The material contains the major elements from the TiO_2/CoO/Ag nanoscale architecture. Such studies provide useful information about filling NTs with another layer of a different material and smaller nanoparticles. When either Ni or Co particles covered by Bi-Te are used, the major difference is that the magnetic field applied can facilitate the transfer of the nanoparticles from solution to the TiO_2 NTs due to the strong interaction of Ni or Co in the magnetic field.

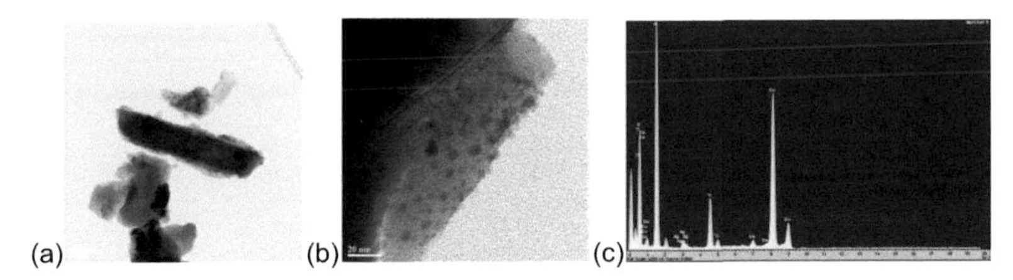

Figure 10.8 TEM observation and element analysis of nanocast TiO_2/Ag/CoO nanofibers or nanorods: (a) TiO_2/Ag/CoO nanorods, (b) Ag nanoparticles in a nanorod, and (c) EDX diffraction spectrum. Republished with permission of PERGAMON, from Ref. [157] (2010); permission conveyed through Copyright Clearance Center, Inc.

The effects of other nanocasting conditions, such as temperature, time, concentrations of aniline, pH values of the solutions in the compositions, and structure uniformity of the nanocomposites, were studied. The prepared composite materials were evaluated by both electron microscopic observation and EDX diffraction spectrum analysis.

10.5 Thermoelectric Property of Nanocast Composites

10.5.1 TiO_2 Nanotube/Polyaniline Polymer Composite

Following the manufacturing process studies, the Seebeck coefficient was measured for the TiO_2 NT/PANI specimens. Before testing, the nanocomposite was lifted off from the Ti substrate. During the Seebeck coefficient measurements, the two ends of the samples were bonded to two separate gold wires with an Ag-based conductive adhesive, which allows high electrical conductivity at the wire/sample interface. Then, one end of the specimens as the hot end was heated up by the Peltier device to a specifically required temperature; the other end, as the cold end, was kept in air at the ambient temperature of 25°C. The hot-end temperature ranges from 40°C to 150°C. The Seebeck coefficient was calculated by the ratio of $\Delta E/\Delta T$. ΔE represents the voltage difference between the hot and cold ends. ΔT is the temperature difference between the two ends.

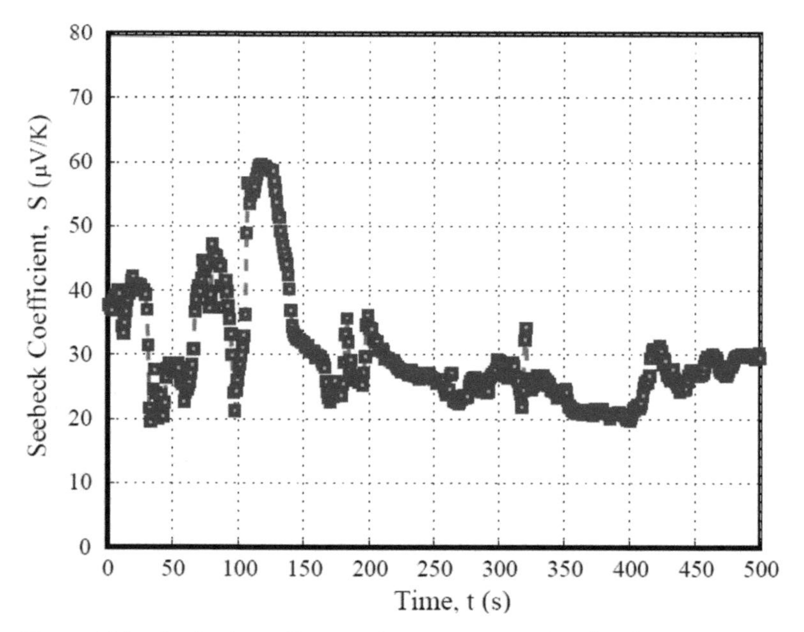

Figure 10.9 Seebeck coefficient of a nanocast polyaniline/TiO_2 nanotube composite. Republished with permission of ASME, from Ref. [165] (2015); permission conveyed through Copyright Clearance Center, Inc.

It is found that the stable absolute value of the Seebeck coefficient for the composite material of PANI nanofibers embedded in TiO_2 NTs is about 30 μV/K, as shown in Fig. 10.9. The material shows n-type behavior. As compared with the inorganic semiconducting material, a bulky silicon crystalline material, the nanofiber has a slightly lower Seebeck coefficient value. The measured Seebeck value for the silicon bulk material is about 40 μV/K under the same measurement condition. Therefore, further improvement of the TE property of the composite material is needed. The electrical resistance of the nanocast PANI nanofiber was measured at room temperature (25°C). The material shows a resistance of 35 MΩ, which is in the range of the resistance showed by typical semiconducting materials.

It is also found that the Seebeck coefficient is time dependent in the initial stage. This is because with the transient heat conduction at the moment of contact, the effective temperature difference fluctuates at the beginning. The heat-induced electron ejection behavior stabilizes after 3 min. Therefore, the measured Seebeck coefficient changes significantly with time at $t < 200$ s. After that, the absolute values of the Seebeck coefficient are stable, as shown in Fig. 10.9.

10.5.2 TiO_2 Nanotube/CoO Ceramic Composite

Ceramic composites were nanocast using AAO as the template following the manufacturing process as described in Ref. [157]. For comparison, the measured Seebeck coefficients at different temperatures are listed in Table 10.1. The four nanocast composite materials are named TiO_2, CoO, TiO_2+Ag, and TiO_2+Ag+CoO in the table. The data show that the nanocast TiO_2+Ag has the highest absolute value of Seebeck coefficient, which may be due to the high electrical conductivity of Ag particles from the electron tunneling between the Ag nanoparticles within the NTs. The nanocast CoO tube composite showed the lowest absolute value of Seebeck coefficient. It is due to the low electrical conductivity. The positive effect of Ag particles was found to be more intensive than the negative effect of the CoO for the Seebeck coefficient. That is why the nanocast TiO_2+Ag+CoO composite has a higher absolute value of Seebeck coefficient than the nanocast TiO_2. The nanocomposites in terms of the absolute values of the Seebeck coefficient are ranked as TiO_2+Ag > TiO_2+Ag+CoO > TiO_2 > CoO.

Table 10.1 Seebeck coefficients of nanocast TiO_2, CoO, TiO_2+Ag, and TiO_2+Ag+CoO nanocomposite materials at different hot-end temperatures, T_h[a]

T_h	TiO_2	CoO	TiO_2+Ag	TiO_2+Ag+CoO
40°C	-220 μV/K	-93 μV/K	-393 μV/K	-300 μV/K
45°C	-210 μV/K	-90 μV/K	-320 μV/K	-255 μV/K
50°C	-172 μV/K	-68 μV/K	-296 μV/K	-208 μV/K
55°C	-153 μV/K	-80 μV/K	-317 μV/K	-200 μV/K
60°C	-126 μV/K	-71 μV/K	-274 μV/K	-163 μV/K
70°C	-107 μV/K	-67 μV/K	-227 μV/K	-167 μV/K
80°C	-98 μV/K	-60 μV/K	-246 μV/K	-130 μV/K
90°C	-102 μV/K	-60 μV/K	-228 μV/K	-114 μV/K
110°C	-79 μV/K	-57 μV/K	-207 μV/K	-108 μV/K
130°C	-91 μV/K	-34 μV/K	-153 μV/K	-108 μV/K

[a]The cold end was kept at an ambient temperature of 25°C

Source: Republished with permission of PERGAMON, from Ref. [157] (2010); permission conveyed through Copyright Clearance Center, Inc.

10.6 Conclusions

Nanocasting as a nontraditional manufacturing technology can be used to make diversified nanocomposite materials. It is shown that a polymer-based composite material (PANI in TiO_2 NT) can be nanocast with the assistance of electric and magnetic fields. The nanocomposite shows n-type property with an average absolute value of Seebeck coefficient of 30 μV/K. The electrical resistance of the composite is about 35 MΩ. It is also concluded that use of nanocasting to make ceramic composite materials shows promise. Metal nanoparticles can also form in situ within composite materials during the nanocasting process. Oxide-based composites containing TiO_2 NTs and TiO_2-CoO coaxial nanocables have a strong Seebeck effect. The absolute value of the Seebeck coefficient is 393 μV/K for the TiO_2 NT. The TiO_2-CoO coaxial nanocable has a slightly lower value, of 300 μV/K. Both composites are n-type. The TE figure of merit of such nanocomposites could potentially be very high due to the low value of the thermal conductivity of the alumina ceramic matrix.

References

1. Shackelford, J. F. (2005). *Introduction to Materials Science for Engineers*, Upper Saddle River, New Jersey: Pearson Prentice Hall, p. 423.
2. Callister, W. D. (2003). *Materials Science and Engineering: An Introduction*, New York: John Wiley & Sons, p. 355.
3. Gan, Y. X., Overfelt, R. A. (2006). Fatigue property of semisolid A357 aluminum alloy under different heat treatment conditions, *J. Mater. Sci.*, **43**, pp. 7537–7544.
4. Attard, G. S., Glyde, J. C., Goltner, C. G. (1995). Liquid-crystalline phases as templates for the synthesis of mesoporous silica, *Nature*, **378**, pp. 366–368.
5. Goltner, C. G., Antonietti, M. (1997). Mesoporous materials by templating of liquid crystalline phases, *Adv. Mater.*, **9**, pp. 431–436.
6. Goltner, C. G., Henke, S., Weibenberger, M. C., Antonietti, M. (1998). Mesoporous silica from lyotropic liquid crystal polymer templates, *Angew. Chem. Int. Ed.*, **110**, pp. 633–636.
7. Kramer, E., Forster, S., Goltner, C., Antonietti, M. (1998). Synthesis of nanoporous silica with new pore morphologies by templating the assemblies of ionic block copolymers, *Langmuir*, **14**, pp. 2027–2031.
8. Goltner, C. G., Berton, B., Kramer, E., Antonietti, M. (1999). Nanoporous silicas by casting the aggregates of amphiphilic block copolymers: the transition from cylinders to lamellae and vesicles, *Adv. Mater.*, **11**, pp. 395–398.
9. Han, B. H., Smarsly, B., Gruber, C., Wenz, G. (2003). Towards porous silica materials via nanocasting of stable pseudopolyrotaxanes from alpha-cyclodextrin and polyamines, *Microporous Mesoporous Mater.*, **66**, pp. 127–132.
10. Xia, X. H., Tu, J. P., Zhang, J., Xiang, J. Y., Wang, X. L., Zhao, X. B. (2010). Cobalt oxide ordered bowl-like array films prepared by electrodeposition through monolayer polystyrene sphere template and electrochromic properties, *ACS Appl. Mater. Interfaces*, **2**, pp. 186–192.
11. Cruz, H. S., Spino, J., Grathwohl, G. (2008). Nanocrystalline ZrO_2 ceramics with idealized macropores, *J. Eur. Ceram. Soc.*, **28**, pp. 1783–1791.

12. Chai, F., Li, D., Wu, H., Zhang, C., Wang, X. (2009). Fabrication of $Cs_{2.5}H_{0.5}PW_{12}O_{40}$ three-dimensional ordered film by colloidal crystal template, *J. Solid State Chem.*, **182**, pp. 1661–1665.
13. Jiang, P., Bertone, J. F., Hwang, K. S., Colvin, V. L. (1999). Single-crystal colloidal multilayers of controlled thickness, *Chem. Mater.*, **11**, pp. 2132–2140.
14. Wang, Z., Li, F., Ergang, N. S., Stein, A. (2006). Effects of hierarchical architecture on electronic and mechanical properties of nanocast monolithic porous carbons and carbon-carbon nanocomposites, *Chem. Mater.*, **18**, pp. 5543–5553.
15. Wang, Z., Li, F., Ergang, N. S., Stein, A. (2008). Synthesis of monolithic 3D ordered macroporous carbon/nano-silicon composites by diiodosilane decomposition, *Carbon*, **46**, pp. 1702–1710.
16. Zhang, F., Wong, S. S. (2009). Controlled synthesis of semiconducting metal sulfide nanowires, *Chem. Mater.*, **21**, pp. 4541–4554.
17. Kumar, A., Whitesides, G. M. (1993). Features of gold having micrometer to centimeter dimensions can be formed through a combination of stamping with an elastomeric stamp and an alkanethiol ink followed by chemical etching, *Appl. Phys. Lett.*, **63**, pp. 2002–2004.
18. Wang, X., Han, G. R. (2003). Fabrication and characterization of anodic aluminum oxide template, *Microelectron. Eng.*, **66**, pp. 166–170.
19. Yu, W. H., Fei, G. T., Chen, X. M., Xue, F. H., Xu, X. J. (2006). Influence of defects on the ordering degree of nanopores made from anodic aluminum oxide, *Phys. Letts. A*, **350**, pp. 392–395.
20. Hu, H., He, D. (2006). Fabrication of Si nanodot arrays by plasma enhanced CVD using porous alumina templates, *Mater. Lett.*, **60**, pp. 1019–1022.
21. Kim, K. T., Cho, S. M. (2006). A simple method for formation of metal nanowires on flexible polymer film, *Mater. Lett.*, **60**, pp. 352–355.
22. Pellin, M. J., Stair, P. C., Xiong, G., Elam, J. W., Birell, J., Curtiss, L., George, S. M., Han, C. Y., Iton, L., Kung, H., Kung, M., Wang, H. H. (2005). Mesoporous catalytic membranes: synthetic control of pore size and wall composition, *Catal. Lett.*, **102**, pp. 127–130.
23. Li, N., Li, X., Yin, X., Wang, W., Qiu, S. (2004). Electroless deposition of open-end Cu nanotube arrays, *Solid State Commun.*, **132**, pp. 841–844.
24. Zhao, Y., Chen, M., Zhang, Y., Xu, T., Liu, W. (2005). A facile approach to formation of through-hole porous anodic aluminum oxide film, *Mater. Lett.*, **59**, pp. 40–43.

25. Hwang, S. K., Jeong, S. W., Lee, O. J., Lee, K. H. (2005). Fabrication of vacuum tube arrays with a sub-micron dimension using anodic aluminum oxide nano-templates, *Microelectron. Eng.*, **77**, pp. 2–7.

26. Wang, X., Wang, X., Huang, W., Sebastian, P. J., Gamboa, S. (2005). Sol-gel template synthesis of highly ordered MnO_2 nanowire arrays, *J. Power Sources*, **140**, pp. 211–215.

27. Kong, L. B. (2005). Synthesis of Y-junction carbon nanotubes within porous anodic aluminum oxide template, *Solid State Commun.*, **133**, pp. 527–529.

28. Chen, P. L., Chang, J. K., Pan, F. M., Kuo, C. T. (2005). Tube number density control of carbon nanotubes on anodic aluminum oxide template, *Diamond Relat. Mater.*, **14**, pp. 804–809.

29. Lee, O. J., Hwang, S. K., Jeong, S. H., Lee, P. S., Lee, K. H. (2005). Synthesis of carbon nanotubes with identical dimensions using an anodic aluminum oxide template on a silicon wafer, *Synth. Met.*, **148**, pp. 263–266.

30. Sun, X. Y., Xu, F. Q., Li, Z. M., Zhang, W. H. (2005). Cyclic voltammetry for the fabrication of high dense silver nanowire arrays with the assistance of AAO template, *Mater. Chem. Phys.*, **90**, pp. 69–72.

31. Yang, Z., Huang, Y., Dong, B., Li, H. L. (2005). Template induced sol-gel synthesis of highly ordered $LaNiO_3$ nanowires, *J. Solid State Chem.*, **178**, pp. 1157–1164.

32. Singh, R. A., Yoon, E. S., Kim, H. J., Kim, J., Jeong, H. E., Suh, K. Y. (2007). Replication of surfaces of natural leaves for enhanced micro-scale tribological property, *Mater. Sci. Eng. C*, **27**, pp. 875–879.

33. Yuan, Z., Chen, H., Zhang, J. (2008). Facile method to prepare lotus-leaf-like super-hydrophobic poly(vinyl chloride) film, *Appl. Surf. Sci.*, **254**, pp. 1593–1598.

34. Lu, A. H., Schuth, F. (2005). Nanocasting pathways to create ordered mesoporous solids, *C. R. Chim.*, **8**, pp. 609–620.

35. Puertolas, B., Solsona, B., Agouram, S., Murillo, R., Mastral, A. M., Aranda, A., Taylor, S. H., Garcia, T. (2010). The catalytic performance of mesoporous cerium oxides prepared through a nanocasting route for the total oxidation of naphthalene, *Appl. Catal. B*, **93**, pp. 395–405.

36. Wu, Z., Meng, Y., Zhao, D. (2010). Nanocasting fabrication of ordered mesoporous phenol-formaldehyde resins with various structures and their adsorption performances for basic organic compounds, *Microporous Mesoporous Mater.*, **128**, pp. 165–179.

37. Lepoutre, S., Smatt, J. H., Laberty, C., Amenitsch, H., Grosso, D., Linden, M. (2009). Detailed study of the pore-filling processes during nanocasting of mesoporous films using SnO_2/SiO_2 as a model system, *Microporous Mesoporous Mater.*, **123**, pp. 185–192.

38. Escax, V., Imperor-Clerc, M., Bazin, D., Davidson, A. (2005). Nanocasting, templated syntheses and structural studies of manganese oxide nanoparticles nucleated in the pores of ordered mesoporous silicas(SBA-15), *C. R. Chim.*, **8**, pp. 663–677.

39. Valdes-Solis, T., Fuertes, A. B. (2006). High-surface area inorganic compounds prepared by nanocasting techniques, *Mater. Res. Bull.*, **41**, pp. 2187–2197.

40. Vantomme, A., Surahy, L., Su, B. L. (2007). Highly ordered mesoporous carbon materials CMI-8 with variable morphologies synthesised by nanocasting, *Colloids Surf. A*, **300**, pp. 65–69.

41. Rigby, S. P., Beanlands, K., Evbuomwan, I. O., Watt-Smith, M. J., Edler, K. J., Fletcher, R. S. (2004). Nanocasting of novel, designer-structured catalyst supports, *Chem. Eng. Sci.*, **59**, pp. 5113–5120.

42. Parmentier, J., Saadhallah, S., Reda, M., Gibot, P., Roux, M., Vidal, L., Vix-Guterl, C., Patarin, J. (2004). New carbons with controlled nanoporosity obtained by nanocasting using a SBA-15 mesoporous silica host matrix and different preparation routes, *J. Phys. Chem. Solids*, **65**, pp. 139–146.

43. Roggenbuck, J., Schafer, H., Tsoncheva, T., Minchev, C., Hanss, J., Tiemann, M. (2007). Mesoporous CeO_2: synthesis by nanocasting, characterisation and catalytic properties, *Microporous Mesoporous Mater.*, **101**, pp. 335–341.

44. Satishkumar, G., Titelman, L., Landau, M. V. (2009). Mechanism for the formation of tin oxide nanoparticles and nanowires inside the mesopores of SBA-15, *J. Solid State Chem.*, **182**, pp. 2822–2828.

45. Xing, R., Rankin, S. E. (2008). Reactive pore expansion during ammonia vapor post-treatment of ordered mesoporous silica prepared with mixed glucopyranoside and cationic surfactants, *Microporous Mesoporous Mater.*, **108**, pp. 65–76.

46. Yan, J., Wang, A., Kim, D. P. (2007). Preparation of ordered mesoporous SiCN ceramics with large surface area and high thermal stability, *Microporous Mesoporous Mater.*, **100**, pp. 128–133.

47. Lebeau, B., Parmentier, J., Soulard, M., Fowler, C., Zana, R., Vix-Guterl, C., Patarin, J. (2005). Organized mesoporous solids: mechanism of formation and use as host materials to prepare carbon and oxide replicas, *C. R. Chim.*, **8**, pp. 597–607.

48. Wang, Y., Wang, Y., Ren, J., Mi, Y., Zhang, F., Li, C., Liu, X., Guo, Y., Guo, Y., Lu, G. (2010). Synthesis of morphology-controllable mesoporous Co_3O_4 and CeO_2, *J. Solid State Chem.*, **183**, pp. 277–284.

49. Sampieri, A., Pronier, S., Blanchard, J., Breysse, M., Brunet, S., Fajerwerg, K., Louis, C., Perot, G. (2005). Hydrodesulfurization of dibenzothiophene on MoS_2/MCM-41 and MoS_2/SBA-15 catalysts prepared by thermal spreading of MoO_3, *Catal. Today*, **107–108**, pp. 537–544.

50. Fajula, F., Galarneau, A., Di Renzo, F. (2005). Advanced porous materials: new developments and emerging trends, *Microporous Mesoporous Mater.*, **82**, pp. 227–239.

51. Zhu, J., Gao, Q., Chen, Z. (2008). Preparation of mesoporous copper cerium bimetal oxides with high performance for catalytic oxidation of carbon monoxide, *Appl. Catal. B*, **81**, pp. 236–243.

52. van der Meer, J., Bardez, I., Bart, F., Albouy, P. A., Wallez, G., Davidson, A. (2009). Dispersion of Co_3O_4 nanoparticles within SBA-15 using alkane solvents, *Microporous Mesoporous Mater.*, **118**, pp. 183–188.

53. Lopes, I., Davidson, A., Thomas, C. (2007). Calibrated Co_3O_4 nanoparticles patterned in SBA-15 silicas: accessibility and activity for CO oxidation, *Catal. Commun.*, **8**, pp. 2105–2109.

54. Roggenbuck, J., Waitz, T., Tiemann, M. (2008). Synthesis of mesoporous metal oxides by structure replication: strategies of impregnating porous matrices with metal salts, *Microporous Mesoporous Mater.*, **113**, pp. 575–582.

55. Kim, Y. S., Guo, X. F., Kim, G. J. (2010). Synthesis of carbon monolith with bimodal meso/macroscopic pore structure and its application in asymmetric catalysis, *Catal. Today*, **150**, pp. 91–99.

56. Xing, W., Zhuo, S. P., Gao, X. (2009). alpha-Fe-incorporated nanoporous carbon with magnetic properties, *Mater. Lett.*, **63**, pp. 1177–1179.

57. Zhu, S., Zhou, Z., Zhang, D., Wang, H. (2006). Synthesis of mesoporous amorphous MnO_2 from SBA-15 via surface modification and ultrasonic waves, *Microporous Mesoporous Mater.*, **95**, pp. 257–264.

58. Fuertes, A. B., Sevilla, M., Alvarez, S., Valdes-Solis, T. (2008). Control of the structural properties of mesoporous polymers synthesized using porous silica materials as templates, *Microporous Mesoporous Mater.*, **112**, pp. 319–326.

59. Lu, A. H., Li, W. C., Schmidt, W., Schuth, F. (2005). Template synthesis of large pore ordered mesoporous carbon, *Microporous Mesoporous Mater.*, **80**, pp. 117–128.

60. Abellan, G., Carrillo, A. I., Linares, N., Serrano, E., Garcia-Martinez, J. (2009). Hierarchical control of porous silica by pH adjustment: alkyl polyamines as surfactants for bimodal silica synthesis and its carbon replica, *J. Solid State Chem.*, **182**, pp. 2141–2148.

61. Pevida, C., Snape, C. E., Drage, T. C. (2009). Templated polymeric materials as adsorbents for the postcombustion capture of CO_2, *Energy Procedia*, **1**, pp. 869–874.

62. Sun, J., Ma, D., Zhang, H., Bao, X., Weinberg, G., Su, D. (2007). Macro-mesoporous silicas complex and the carbon replica, *Microporous Mesoporous Mater.*, **100**, pp. 356–360.

63. Alvarez, S., Valdes-Solis, T., Fuertes, A. B. (2008). Templated synthesis of nanosized mesoporous carbons, *Mater. Res. Bull.*, **43**, pp. 1898–1904.

64. Lu, A. H., Li, W. C., Schmidt, W., Kiefer, W., Schuth, F. (2004). Easy synthesis of an ordered mesoporous carbon with a hexagonally packed tubular structure, *Carbon*, **42**, pp. 2939–2948.

65. Xiang, L., Royer, S., Zhang, H., Tatibouet, J. M., Barrault, J., Valange, S. (2009). Properties of iron-based mesoporous silica for the CWPO of phenol: a comparison between impregnation and co-condensation routes, *J. Hazard. Mater.*, **172**, pp. 1175–1184.

66. Goltner-Spickermann, C. (2002). Non-ionic templating of silica: formation mechanism and structure, *Curr. Opin. Colloid Interface Sci.*, **7**, pp. 173–178.

67. Yue, W., Zhou, W. (2008). Crystalline mesoporous metal oxide, *Prog. Nat. Sci.*, **18**, pp. 1329–1338.

68. Wagner, T., Sauerwald, T., Kohl, C. D., Waitz, T., Weidmann, C., Tiemann, M. (2009). Gas sensor based on ordered mesoporous In_2O_3, *Thin Solid Films*, **517**, pp. 6170–6175.

69. de Lima, R. K. C., Batista, M. S., Wallau, M., Sanches, E. A., Mascarenhas, Y. P., Urquieta-Gonzalez, E. A. (2009). High specific surface area LaFeCo perovskites-Synthesis by nanocasting and catalytic behavior in the reduction of NO with CO, *Appl. Catal. B*, **90**, pp. 441–450.

70. Lu, A. H., Tuysuz, H., Schuth, F. (2008). Synthesis of ordered mesoporous carbon containing highly dispersed copper-sulphur compounds in the carbon framework via a nanocasting route, *Microporous Mesoporous Mater.*, **111**, pp. 117–123.

71. Lai, X., Wang, H., Mao, D., Yang, N., Yao, J., Xing, C., Wang, D., Li, X. (2008). Mesoporous indium oxide synthesized via a nanocasting route, *Mater. Lett.*, **62**, pp. 3868–3871.

72. Wu, D., Liang, Y., Yang, X., Li, Z., Zou, C., Zeng, X., Lu, G., Fu, R. (2008). Direct fabrication of bimodal mesoporous carbon by nanocasting, *Microporous Mesoporous Mater.*, **116**, pp. 91–94.

73. Pacula, A., Mokaya, R. (2007). Layered double hydroxides as templates for nanocasting porous N-doped graphitic carbons via chemical vapour deposition, *Microporous Mesoporous Mater.*, **106**, pp. 147–154.

74. Li, S., Liang, Y., Wu, D., Fu, R. (2010). Fabrication of bimodal mesoporous carbons from petroleum pitch by a one-step nanocasting method, *Carbon*, **48**, pp. 839–843.

75. Bai, P., Xing, W., Wu, P., Liu, X., Yan, Z., Zhao, X. S. (2007). Evaporation-controlled nanocasting approach to a precision replication at nanometer scale, *Mater. Lett.*, **61**, pp. 4231–4234.

76. Nakamura, T., Yamada, Y., Yano, K. (2009). Monodispersed nanoporous starburst carbon spheres and their three-dimensionally ordered arrays, *Microporous Mesoporous Mater.*, **117**, pp. 478–485.

77. Xu, L. Y., Shi, Z. G., Feng, Y. Q. (2007). A facile route to a silica monolith with ordered mesopores and tunable through pores by using hydrophilic urea formaldehyde resin as a template, *Microporous Mesoporous Mater.*, **98**, pp. 303–308.

78. Perez-Cabero, M., Esteve-Turrillas, F. A., Beltran, D., Amoros, P. (2010). Hierarchical porous carbon with designed pore architecture and study of its adsorptive properties, *Solid State Sci.*, **12**, pp. 15–25.

79. Lu, A. H., Smatt, J. H., Backlund, S., Linden, M. (2004). Easy and flexible preparation of nanocasted carbon monoliths exhibiting a multimodal hierarchical porosity, *Microporous Mesoporous Mater.*, **72**, pp. 59–65.

80. Dibandjo, P., Chassagneux, F., Bois, L., Sigala, C., Miele, P. (2006). Synthesis of boron nitride with a cubic mesostructure, *Microporous Mesoporous Mater.*, **92**, pp. 286–291.

81. Liu, Q., Wang, A., Xu, J., Zhang, Y., Wang, X., Zhang, T. (2008). Preparation of ordered mesoporous crystalline alumina replicated by mesoporous carbon, *Microporous Mesoporous Mater.*, **116**, pp. 461–468.

82. Li, H., Zhu, S., Xi, H., Wang, R. (2006). Nickel oxide nanocrystallites within the wall of ordered mesoporous carbon CMK-3: synthesis and characterization, *Microporous Mesoporous Mater.*, **89**, pp. 196–203.

83. Ania, C. O., Pernak, J., Stefaniak, F., Raymundo-Pinero, E., Beguin, F. (2009). Polarization-induced distortion of ions in the pores of carbon electrodes for electrochemical capacitors, *Carbon*, **47**, pp. 3158–3166.

84. Yang, Y. X., Bourgeois, L., Zhao, C., Zhao, D., Chaffee, A., Webley, P. A. (2009). Ordered micro-porous carbon molecular sieves containing

well-dispersed platinum nanoparticles for hydrogen storage, *Microporous Mesoporous Mater.*, **119**, pp. 39–46.

85. Zhou, L., Li, H., Yu, C., Zhou, X., Tang, J., Meng, Y., Xia, Y., Zhao, D. (2006). Easy synthesis and supercapacities of highly ordered mesoporous polyacenes/carbons, *Carbon*, **44**, pp. 1581–1616.

86. Jaroniec, M., Gorka, J., Choma, J., Zawislak, A. (2009). Synthesis and properties of mesoporous carbons with high loadings of inorganic species, *Carbon*, **47**, pp. 3034–3040.

87. Zhou, J., He, J., Zhang, C., Wang, T., Sun, D., Di, Z., Wang, D. (2010). Mesoporous carbon spheres with uniformly penetrating channels and their use as a supercapacitor electrode material, *Mater. Charact.*, **61**, pp. 31–38.

88. Popp, A., Engstler, J., Schneider, J. J. (2009). Porous carbon nanotube-reinforced metals and ceramics via a double templating approach, *Carbon*, **47**, pp. 3208–3214.

89. Perez-Cabero, M., Puchol, V., Beltran, D., Amoros, P. (2008). Thalassiosira pseudonana diatom as biotemplate to produce a macroporous ordered carbon-rich material, *Carbon*, **46**, pp. 297–304.

90. Athens, G. L., Shayib, R. M., Chmelka, B. F. (2009). Functionalization of mesostructured inorganic-organic and porous inorganic materials, *Curr. Opin. Colloid Interface Sci.*, **14**, pp. 281–292.

91. Li, H., Xi, H., Zhu, S., Wang, R. (2006). Electrochemical lithium storage of Li-Ti-O compound calcined at different temperatures, *Mater. Lett.*, **60**, pp. 943–946.

92. Kim, J. H., Fang, B., Kim, M., Yu, J. S. (2009). Hollow spherical carbon with mesoporous shell as a superb anode catalyst support in proton exchange membrane fuel cell, *Catal. Today*, **146**, pp. 25–30.

93. Srinivasu, P., Vinu, A., Hishita, S., Sasaki, T., Ariga, K., Mori, T. (2008). Preparation and characterization of novel microporous carbon nitride with very high surface area via nanocasting technique, *Microporous Mesoporous Mater.*, **108**, pp. 340–344.

94. Parmentier, J., Valtchev, V., Gaslain, F., Tosheva, L., Ducrot-Boisgontier, C., Moller, J., Patarin, J., Vix-Guterl, C. (2009). Effect of the zeolite crystal size on the structure and properties of carbon replicas made by a nanocasting process, *Carbon*, **47**, pp. 1066–1073.

95. Ducrot-Boisgontier, C., Parmentier, J., Patarin, J. (2009). Influence of the carbon precursors on the structural properties of EMT-type nanocasted-carbon replicas, *Microporous Mesoporous Mater.*, **126**, pp. 101–106.

96. Fang, Y., Hu, H., Chen, G. (2008). Zeolite with tunable intracrystal mesoporosity synthesized with carbon aerogel as a secondary template, *Microporous Mesoporous Mater.*, **113**, pp. 481–489.

97. Deshpande, A. S., Niederberger, M. (2007). Synthesis of mesoporous ceria zirconia beads, *Microporous Mesoporous Mater.*, **101**, pp. 413–418.

98. Shang, M., Wang, W., Zhou, L., Sun, S., Yin, W. (2009). Nanosized $BiVO_4$ with high visible-light-induced photocatalytic activity: ultrasonic-assisted synthesis and protective effect of surfactant, *J. Hazard. Mater.*, **172**, pp. 338–344.

99. Xing, R., Rankin, S. E. (2007). Use of the ternary phase diagram of a mixed cationic/glucopyranoside surfactant system to predict mesostructured silica synthesis, *J. Colloid Interface Sci.*, **316**, pp. 930–938.

100. Sogo, K., Nakajima, M., Kawata, H., Hirai, Y. (2007). Reproduction of fine structures by nanocasting lithography, *Microelectron. Eng.*, **84**, pp. 909–911.

101. Hirai, Y., Yoshikawa, T., Morimatsu, M., Nakajima, M., Kawata, H. (2005). Fine pattern transfer by nanocasting lithography, *Microelectron. Eng.*, **78–79**, pp. 641–646.

102. Fang, Y., Hu, H. (2007). Mesoporous TS-1: nanocasting synthesis with CMK-3 as template and its performance in catalytic oxidation of aromatic thiophene, *Catal. Commun.*, **8**, pp. 817–820.

103. Rao, M. P., Goldberg-Oppenheimer, P., Kababya, S., Vegab, S., Landau, M. V. (2007). Proton enriched high-surface-area cesium salt of phosphotungstic heteropolyacid with enhanced catalytic activity fabricated by nanocasting strategy, *J. Mol. Catal. A: Chem.*, **275**, pp. 214–227.

104. Valdes-Solis, T., Valle-Vigon, P., Alvarez, S., Marban, G., Fuertes, A. B. (2007). Manganese ferrite nanoparticles synthesized through a nanocasting route as a highly active Fenton catalyst, *Catal. Commun.*, **8**, pp. 2037–2042.

105. Zhang, H., Yan, X., Li, W. (2009). Nanocast ordered mesoporous CeO_2 as support for highly active gold catalyst in CO oxidation, *Chin. J. Catal.*, **30**, pp. 1085–1090.

106. Zhu, J., Gao, Q. (2009). Mesoporous MCo_2O_4 (M = Cu, Mn and Ni) spinels: structural replication, characterization and catalytic application in CO oxidation, *Microporous Mesoporous Mater.*, **124**, pp. 144–152.

107. Rahman, M. S., Rankin, S. E. (2010). Predictive synthesis of ordered mesoporous silica with maltoside and cationic surfactants based on aqueous lyotropic phase behavior, *J. Colloid Interface Sci.*, **342**, pp. 33–42.

108. Abecassis-Wolfovich, M., Landau, M. V., Brenner, A., Herskowitz, M. (2008). Low-temperature combustion of 2,4,6-trichlorophenol in catalytic wet oxidation with nanocasted Mn–Ce-oxide catalyst, *J. Catal.*, **247**, pp. 201–213.

109. Joo, S. H., Lee, H. I., You, D. J., Kwon, K., Kim, J. H., Choi, Y. S., Kang, M., Kim, J. M., Pak, C., Chang, H., Seung, D. (2008). Ordered mesoporous carbons with controlled particle sizes as catalyst supports for direct methanol fuel cell cathodes, *Carbon*, **46**, pp. 2034–2045.

110. Chai, G. S., Yoon, S. B., Yu, J. S. (2005). Highly efficient anode electrode materials for direct methanol fuel cell prepared with ordered and disordered arrays of carbon nanofibers, *Carbon*, **43**, pp. 3028–3031.

111. Smatt, J. H., Schuwer, N., Jarn, M., Lindner, W., Linden, M. (2008). Synthesis of micrometer sized mesoporous metal oxide spheres by nanocasting, *Microporous Mesoporous Mater.*, **112**, pp. 308–318.

112. Walcarius, A., Kuhn, A. (2008). Ordered porous thin films in electrochemical analysis, *Trends Anal. Chem.*, **27**, pp. 593–603.

113. Leitner, A., Sturma, M., Smatt, J. H., Jarn, M., Linden, M., Mechtler, K., Lindner. W. (2009). Optimizing the performance of tin dioxide microspheres for phosphopeptide enrichment, *Anal. Chim. Acta*, **638**, pp. 51–57.

114. Soler-Illia, G. J. A. A., Crepaldi, E. L., Grosso, D., Sanchez, C. (2003). Block copolymer-templated mesoporous oxides, *Curr. Opin. Colloid Interface Sci.*, **8**, pp. 109–126.

115. Li, Z., Jia, Z., Luan, Y., Mu, T. (2008). Ionic liquids for synthesis of inorganic nanomaterials, *Curr. Opinion Solid State Mater. Sci.*, **12**, pp. 1–8.

116. Wan, L., Fu, H., Shi, K., Tian, X. (2008). Facile synthesis of ordered nanocrystalline alumina thin films with tunable mesopore structures, *Microporous Mesoporous Mater.*, **115**, pp. 301–307.

117. Chen, X., Kong, L., Dong, D., Yang, G., Yu, L., Chen, J., Zhang, P. (2009). Synthesis and characterization of superhydrophobic functionalized $Cu(OH)_2$ nanotube arrays on copper foil, *Appl. Surf. Sci.*, **255**, pp. 4015–4019.

118. Ma, Y., Qi, L. (2009). Solution-phase synthesis of inorganic hollow structures by templating strategies, *J. Colloid Interface Sci.*, **335**, pp. 1–10.

119. Calvillo, L., Celorrio, V., Moliner, R., Cabot, P. L., Esparbe, I., Lazaro, M. J. (2008). Control of textural properties of ordered mesoporous materials, *Microporous Mesoporous Mater.*, **116**, pp. 292–298.

120. Li, X., Nandhakumar, I. S., Attard, G. S., Markham, M. L., Smith, D. C., Baumberg, J. J. (2009). Nanotemplated lead telluride thin films, *Microporous Mesoporous Mater.*, **118**, pp. 403–407.

121. Rao, K. K., Hall, D. O. (1984). Photosynthetic production of fuels and chemicals in immobilized systems, *Trends Biotechnol.*, **2**, pp. 124–129.

122. Gassanova, L. G., Netrusov, A. I., Teplyakov, V. V., Modigell, M. (2006). Fuel gases from organic wastes using membrane bioreactors, *Desalination*, **198**, pp. 56–66.

123. Weaver, P. F., Lien, S., Seibert, M. (1980). Photobiological production of hydrogen, *Sol. Energy*, **24**, pp. 3–45.

124. Benemann, J. R., Miyamoto, K., Hallenbeck, P. C. (1980). Bioengineering aspects of biophotolysis, *Enzyme Microb. Technol.*, **2**, pp. 103–111.

125. Miura, Y. (1995). Hydrogen production by biophotolysis based on microalgal photosynthesis, *Process Biochem.*, **30**, pp. 1–7.

126. Lambert, G. R., Smith, G. D. (1977). Hydrogen formation by marine blue—green algae, *FEBS Lett.*, **83**, pp. 159–162.

127. Modigell, M., Holle, N. (1998). Reactor development for a biosolar hydrogen production process, *Renewable Energy*, **14**, pp. 421–426.

128. Kaneko, M., Ueno, H., Saito, R., Yamaguchi, S., Fujii, Y., Nemoto, J. (2009). UV light-activated decomposition/cleaning of concentrated biomass wastes involving also solid suspensions with remarkably high quantum efficiency, *Appl. Catal. B*, **91**, pp. 254–261.

129. Kaneko, M., Ueno, H., Saito, R., Nemoto, J. (2009). Highly efficient photoelectrocatalytic decomposition of biomass compounds using a nanoporous semiconductor photoanode and an O_2-reducing cathode with quantum efficiency over 100, *Catal. Lett.*, **131**, pp. 184–188.

130. Kaneko, M., Ueno, H., Saito, R., Suzuki, S., Nemoto, J., Fujii, Y. (2009). Biophotochemical cell (BPCC) to photodecompose biomass and bio-related compounds by UV irradiation with simultaneous electrical power generation, *J. Photochem. Photobiol. A*, **205**, pp. 168–172.

131. Antoniadou, M., Lianos, P. (2009). Near ultraviolet and visible light photoelectrochemical degradation of organic substances producing electricity and hydrogen, *J. Photochem. Photobiol. A*, **204**, pp. 69–74.

132. Antoniadou, M., Bouras, P., Strataki, N., Lianos, P. (2008). Hydrogen and electricity generation by photoelectrochemical decomposition of

ethanol over nanocrystalline titania, *Int. J. Hydrogen Energy*, **33**, pp. 5045–5051.

133. Ueno, H., Nemoto, J., Ohnuki, K., Horikawa, M., Hoshino, M., Kaneko, M. (2009). Photoelectrochemical reaction of biomass-related compounds in a biophotochemical cell comprising a nanoporous TiO_2 film photoanode and an O_2-reducing cathode, *J. Appl. Electrochem.*, **39**, pp. 1897–1905.

134. Madhavaram, R., Sander, J., Gan, Y. X., Masiulaniec, C. K. (2009). Thermoelectric property of PbTe coating on copper and nickel, *Mater. Chem. Phys.*, **118**, pp. 165–173.

135. Gan, Y. X., Sweetman, J., Lawrence, J. G. (2010). Electrodeposition and morphology analysis of Bi–Te thermoelectric alloy nanoparticles on copper substrate, *Mater. Lett.*, **64**, pp. 449–452.

136. Bao, S. J., Li, C. M., Guo, C. X., Qiao, Y. (2008). Biomolecule-assisted synthesis of cobalt sulfide nanowires for application in supercapacitors, *J. Power Sources*, **180**, pp. 676–681.

137. de Bont, P. W., Vissenberg, M. J., Hensen, E. J. M., de Beer, V. H. J., van Veen, J. A. R., van Santen, R. A., van der Kraan, A. M. (2002). Cobalt–molybdenum-sulfide particles inside NaY zeolite?: a MES and EXAFS study, *Appl. Catal. A*, **236**, pp. 205–222.

138. Wang, J., Ng, S. H., Wang, G. X., Chen, J., Zhao, L., Chen, Y., Liu, H. K. (2006). Synthesis and characterization of nanosize cobalt sulfide for rechargeable lithium batteries, *J. Power Sources*, **159**, pp. 287–290.

139. Kadono, T., Kubota, T., Hiromitsu, I., Okamoto, Y. (2006). Characterization of highly dispersed cobalt sulfide catalysts by X-ray absorption fine structure and magnetic properties, *Appl. Catal. A*, **312**, pp. 125–133.

140. Zhang, Z., Ma, J., Yang, X. (2003). Separate/simultaneous catalytic reduction of sulfur dioxide and/or nitric oxide by carbon monoxide over TiO_2-promoted cobalt sulfides, *J. Mol. Catal. A: Chem.*, **195**, pp. 189–200.

141. Morgenstern, K., Legsgaard, E., Besenbacher, F. (2008). Sintering of cobalt nanoclusters on Ag(1 1 1) by sulfur: formation of one-, two-, and three-dimensional cobalt sulfide, *Surf. Sci.*, **602**, pp. 661–670.

142. Sanders, A. F. H., de Jong, A. M., de Beer, V. H. J., van Veen, J. A. R., Niemantsverdriet, J. W. (1999). Formation of cobalt–molybdenum sulfides in hydrotreating catalysts: a surface science approach, *Appl. Surf. Sci.*, **144–145**, pp. 380–384.

143. Sohrabnezhad, S., Pourahmad, A., Sadjadi, M. S., Sadeghi, B. (2008). Nickel cobalt sulfide nanoparticles grown on AlMCM-41 molecular sieve, *Physica E*, **40**, pp. 684–688.

144. Trikalitis, P. N., Kerr, T. A., Kanatzidis, M. G. (2006). Mesostructured cobalt and nickel molybdenum sulfides, *Microporous Mesoporous Mater.*, **88**, pp. 187–190.

145. Choi, J. S., Mauge, F., Pichon, C., Olivier-Fourcade, J., Jumas, J. C., Petit-Clair, C., Uzio, D. (2004). Alumina-supported cobalt–molybdenum sulfide modified by tin via surface organometallic chemistry: application to the simultaneous hydrodesulfurization of thiophenic compounds and the hydrogenation of olefins, *Appl. Catal. A*, **267**, pp. 203–216.

146. Okamoto, Y. (1997). Preparation and characterization of zeolite-supported molybdenum and cobalt-molybdenum sulfide catalysts, *Catal. Today*, **39**, pp. 45–59.

147. Eze, F. C., Okeke, C. E. (1997). Chemical-bath-deposited cobalt sulphide films: preparation effects, *Mater. Chem. Phys.*, **47**, pp. 31–36.

148. Hu, Q. R., Wang, S. L., Zhang, Y., Tang, W. H. (2010). Synthesis of cobalt sulfide nanostructures by a facile solvothermal growth process, *J. Alloys Compd.*, **491**, pp. 707–711.

149. Liu, X. (2005). Hydrothermal synthesis and characterization of nickel and cobalt sulfides nanocrystallines, *Mater. Sci. Eng. B*, **119**, pp. 19–24.

150. Lelias, M. A., Le Guludec, E., Mariey, L., van Gestel, J., Travert, A., Oliviero, L., Mauge, F. (2010). Effect of EDTA addition on the structure and activity of the active phase of cobalt–molybdenum sulfide hydrotreatment catalysts, *Catal. Today*, **150**, pp. 179–185.

151. Vissenberg, M. J., de Bont, P. W., Gruijters, W., de Beer, V. H. J., van der Kraan, A. M., van Santen, R. A., van Veen, J. A. R. (2000). Zeolite Y-supported cobalt sulfide hydrotreating catalysts: III. Prevention of protolysis and the effect of protons on the HDS activity, *J. Catal.*, **189**, pp. 209–220.

152. Rodriguez-Castellon, E., Jimenez-Lopez, A., Eliche-Quesada, D. (2008). Nickel and cobalt promoted tungsten and molybdenum sulfide mesoporous catalysts for hydrodesulfurization, *Fuel*, **87**, pp. 1195–1206.

153. Qian, X. F., Zhang, X. M., Wang, C., Tang, K. B., Xie, Y., Qian, Y. T. (1998). Solvent–thermal preparation of nanocrystalline pyrite cobalt disulfide, *J. Alloys Compd.*, **278**, pp. 110–112.

154. Mijoin, J., Thevenin, V., Garcia Aguirre, N., Yuze, H., Wang, J., Li, W. Z., Perot, G., Lemberton, J. L. (1999). Thioreduction of cyclopentanone and hydrodesulfurization of dibenzothiophene over sulfided nickel or cobalt-promoted molybdenum on alumina catalysts, *Appl. Catal. A*, **180**(1–2), pp. 95–104.

155. Toba, M., Miki, Y., Kanda, Y., Matsui, T., Harada, M., Yoshimura, Y. (2005). Selective hydrodesulfurization of FCC gasoline over $CoMo/Al_2O_3$ sulfide catalyst, *Catal. Today*, **104**, pp. 64–69.

156. Ledoux, M. J., Peter, A., Blekkan, E. A., Luck, F. (1995). The role of the nature and the purity of the alumina support on the hydrodesulfurization activity of CoMo sulfides, *Appl. Catal. A*, **133**(2), pp. 321–333.

157. Su, L., Gan, Y. X., Zhang, L. (2011). Thermoelectricity of nanocomposites containing TiO_2-CoO coaxial nanocables, *Scr. Mater.*, **64**, pp. 745–748.

158. Amalorpava Mary, L., Senthilram, T., Suganya, S., Nagarajan, L., Venugopal, J., Ramakrishna, S., Giri Dev, V. R. (2013). Centrifugal spun ultrafine fibrous web as a potential drug delivery vehicle, *eXPRESS Polym. Lett.*, **7**(3), pp. 238–248.

159. Bao, N., Wei, Z., Ma, Z., Liu, F., Yin, G. (2010). Si-doped mesoporous TiO_2 continuous fibers: preparation by centrifugal spinning and photocatalytic properties, *J. Hazard. Mater.*, **174**, pp. 129–136.

160. Wang, L., Shi, J., Liu, L., Secret, E., Chen, Y. (2011). Fabrication of polymer fiber scaffolds by centrifugal spinning for cell culture studies, *Microelectron. Eng.*, **88**, pp. 1718–1721.

161. Sedagha, A., Taheri-Nassaj, A., Naghizadeh, R. (2006). An alumina mat with a nano microstructure prepared by centrifugal spinning method, *J. Non-Cryst. Solids*, **352**, pp. 2818–2828.

162. Sandou, T., Oya, A. (2005). Preparation of carbon nanotubes by centrifugal spinning of coreshell polymer particles, *Carbon*, **43**, pp. 2013–2032.

163. Suh, H., Jung, H., Hangarter, C. M., Park, H., Lee, Y., Choa, Y., Myung, N. V., Hong, K. (2012). Diameter and composition modulated bismuth telluride nanowires by galvanic displacement reaction of segmented NiFe nanowires, *Electrochim. Acta*, **75**, pp. 201–207.

164. Suh, H., Nam, K. H., Jung, H., Kim, C. Y., Kim, J. G., Kim, C. S., Myung, N. V., Hong, K. (2013). Tapered BiTe nanowires synthesis by galvanic displacement reaction of compositionally modulated NiFe nanowires, *Electrochim. Acta*, **90**, pp. 582–588.

165. Decker, B. Y., Gan, Y. X. (2015). Electric field-assisted additive manufacturing polyaniline based composites for thermoelectric energy conversion, *J. Manuf. Sci. Eng.*, **137**(2), pp. 024504-1–024504-3.

166. Thompson, J. M. T., Stewart, H. B. (1986). *Nonliear Dynamics and Chaos, Geometrical Methods for Engineers and Scientists*, New York: John Wiley and Sons, pp. 51–83.

167. Tabor, M. (1989). *Chaos and Integrability in Nonlinear Dynamics*, New York: John Wiley and Sons, pp. 1–41.

168. Wiggins, S. (1992). *Chaotic Transport in Dynamical Systems*, New York: Springer-Verlag, pp. 17–79.

Chapter 11

Electrohydrodynamic Manufacturing of Thermoelectric Composite Materials

This chapter deals with additive manufacturing of thermoelectric energy conversion functional composite materials by integrating electrohydrodynamic manufacturing into 3D printing. The chapter resolves several major issues related to (i) how to design and make a new manufacturing machine consisting of an electrohydrodynamic manufacturing system and a 3D printer, (ii) how to use the machine to manufacture composite materials with functional thermoelectric nanoparticles uniformly distributed in polymer fibers, and (iii) how to heat-treat the composites to convert the polymer fibers into partially carbonized fibers. In addition, the composite materials with a significant thermoelectric response for application in thermal sensing and energy conversion are discussed. Finally, scalable manufacturing of composite material mats by this new technology is introduced.

11.1 Introduction

Traditional electrohydrodynamic (EHD) manufacturing machines can only produce randomly oriented nanofibers. To control the fiber orientation for practical applications in optoelectronics, energy conversion, sensing, and guided tissue regeneration, a new

Nanomaterials for Thermoelectric Devices
Yong X. Gan

ISBN 978-981-4774-98-7 (Hardcover), 978-0-429-48872-6 (eBook)
www.panstanford.com

manufacturing technology has to be researched. The existing 3D printing technology allows materials to be placed in controlled ways. However, the resolution of 3D printing is limited. Nanofibers can hardly be made directly through 3D printing. Integrating EHD casting into 3D printing results in a new additive manufacturing technology that allows the production of composite materials containing well-aligned 2D and/or 3D nanofibers for special applications. On the basis of our recent experimental research, a new manufacturing machine has been designed by assembling a self-built EHD processing unit in a 3D printer. This EHD processing unit can be driven by the 3D printer to generate preprogrammed *x-y-z* three-directional motions so that the nanofibers produced by the EHD processing unit can be placed through fully controlled preset programs. Fundamental studies has been carried out to understand the science underpinning the new manufacturing technology. The objective of the chapter is to introduce new knowledge of preparing 2D and 3D nanostructured composite materials through a layer-by-layer additive manufacturing process under the action of EHD forces. The nanostructured composite materials made by the new manufacturing technology have high surface areas and enhanced surface activities. Such properties could significantly increase the thermoelectric (TE) sensitivity of the materials to external signals and enhance the phonon-induced electron-hole pair generation at the surface of the materials.

Manufacturing composite materials containing nanofibers and nanoparticles via an innovative process in which EHD casting is integrated into 3D printing has caught much attention. EHD casting works by exposing a small jet of the selected material to a relatively high voltage, usually in the range of 5 kV to 30 kV. This voltage causes the material to undergo stretching and bending to form nanofibers as it gets farther away from the jet. The traditional EHD manufacturing process can only produce randomly oriented nanofibers. To control the fiber orientation for practical applications in optoelectronics, energy conversion, sensing, and guided tissue regeneration, a new manufacturing technology of combining 3D printing and EHD has developed. The existing 3D printing technology allows materials to be placed in controlled ways. However, the resolution of 3D printing is limited. Nanofibers can hardly be made directly through 3D printing. Integrating EHD casting into 3D printing results in a new additive manufacturing technology, which allows the production of composite materials containing well-aligned 2D and/or 3D nanofibers for special applications.

This chapter presents the following aspects to readers: (i) how to design and make a new manufacturing machine consisting of an EHD casting system and a 3D printer, (ii) how to use the machine to manufacture composite materials with functional nanoparticles uniformly distributed in polymer fibers, (iii) how to perform heat treatment on the nanofiber composites to convert the polymer fibers into partially carbonized fibers, (iv) how to test the functions of the composite materials for sensing and energy conversions, and (v) how to characterize the micro- and nanostructures of composite materials to understand the structure-property relations.

EHD forces include electric repulsion, fluid pump pressure, and body force (gravity). Such forces may facilitate a uniform distribution of nanoparticles in polymer nanofibers. The manufactured composite materials may have enhanced TE and photovoltaic properties due to the special structure formation in the combined EHD casting and 3D printing processes. It is possible to make high-performance sensors and energy converters with multiple components and 3D printed architectures. To show this technology, materials consisting of different dispersed nanoparticles in polyacrylonitrile (PAN) nanofibers are made. The nanoparticles are made from bismuth telluride and/or antimony telluride alloy compounds. Both are intrinsic narrow-band semiconductors. Therefore, these nanoparticles can generate TE and photovoltaic functions. The PAN polymer is mixed with the Bi-Te and Sb-Te alloy nanoparticles in an *N,N*-dimethylformamide (DMF) solvent. Uniform dispersion of nanoparticles in polymer nanofibers to form composites can be achieved via combined EHD casting and 3D printing. Then high-temperature heat treatment on the composites in hydrogen gas is conducted to convert the polymer nanofibers into partially carbonized fibers. The partially carbonized fibers are expected to have tuned electrical conductivities depending on the heat treatment temperatures. The higher the heat treatment temperature, the better is the conductivity of the carbon fibers. The carbon fibers will serve as electrical connectors to the Bi-Te or Sb-Te nanoparticles.

The Seebeck coefficient of the composite materials was measured. Since such nanocomposites have low-dimensional structures with high interface areas for phonon scattering, they are expected to have enhanced TE properties, that is high Seebeck coefficients, high electrical conductivity, and low thermal conductivity, which is required for effective energy conversions and sensing. Morphology analysis was performed to understand the effects of some important manufacturing parameters, such as nanoparticle content, polymer concentration, applied EHD casting voltage, pump pressure, 3D printing speed, and heat treatment temperature, on the structure development of the composite materials. The structure-property relation of the composite materials will also be discussed in the chapter.

EHD casting can be demonstrated by exposing a small jet of the desired material to a relatively high DC voltage, usually in the range of several kilovolts to tens of kilovolts. The DC voltage causes the material to undergo stretching and bending as it gets farther away from the voltage point in the pattern of a Taylor cone [1]. EHD casting [2] allows the size of a droplet or a jet from a fluid much smaller than the size of the injection needle. This is because the fluid to be cast is electrified and forms a Taylor cone. Fine droplets or jets then come out from the Taylor cone under the action of the electric force. This new casting method can eliminate the resolution limit from the size of the needle for casting. There is another advantage, that of low cost. The EHD process has made the production of nanoscale fibers cheaper and easier to accomplish than other developed methods, such as chemical vapor deposition [3], X-ray lithography [4], and electron-beam writing followed by ion etching [5]. Therefore, the EHD process has been used for making nanofibers [6], particle-connected-by-nanofiber composites [7], thin wires [8], nanoparticles [9], and microscale metal patterns [10].

In addition to the single-jet EHD, coaxial EHD has been studied for making nanofibers [11] and nanoscale spherical capsules [12]. During the coaxial EHD process, the outer liquid surrounds and encapsulates the inner one, which allows core-shell biodegradable nanofiber made for hydrogen storage [13], tissue repair [11], sustained drug release [14, 15], and biochemical sensing [16]. Besides the biodegradable polymers, various other core and shell materials were used [17]. The core and shell materials could be miscible or immiscible [18], depending on the structure to be made. Among the various core-shell materials, PAN has caught much attention because it can be easily processed into submicron-sized fibers [19], serving as the precursor for carbon nanofibers (CNFs) [20].

Controlling the structure in 2D and 3D forms has caught great interest [21]. For example, inkjet printing has been used for scaffold generation [22]. Although microfabrication technology was used for cell culture formation [23], the more commonly used practice is 3D printing for generating biostructure tissues [24, 25]. Combining 3D printing with other technologies, for example, electrospraying, has recently been studied due to better controllability of the deposition [26]. A combined process such as this has been used for p-type conductive transparent oxide structure preparation [27]. 3D printing is frequently used for food fabrication [28] and biofabrication [29–36]. Although 3D printing can spatially control the material deposition very well, it has limited resolution, that is at a scale of millimeters or several hundreds of micrometers. For nanoscale resolution, EHD processing is the best [37]. EHD processing, including electrospinning and electrospraying, is suitable for generating the required hierarchical scaffolds, for example, micro- and nanomotifs in bone structure [38]. EHD processing has also found application in neural tissue regeneration [39] and functional gradient nanocomposite preparation [40].

One of the ways to preserve the controllability of 3D printing while keep the nanoscale feature formation capability is to develop a combined additive manufacturing process [41]. In view of this trend, a hybrid printing approach using multiple printing heads, including inkjet printing head and electrospinning head, has been proposed [42]. It may be able to partially address the issue on porosity control in cell structure printing [43]. It can also achieve a better resolution of printed features that is normally impossible to get from the traditional rapid prototyping process [44]. Neural tissue generation as discussed in Ref. [45] could potentially be improved. Nutrient mixing and diffusion in tissue structure as mentioned in Ref. [46] can be controlled better by depositing the tailored pore architectures. Other related progress in this direction includes indirect 3D printing [47] and modified 3D writing [48]. Both approaches can generate the nanoscale features as shown by electrospinning [49]. Considering the promising development in this direction, it is a major objective of the chapter to introduce better controllability of the nanostructures by integrating near-field EHD processing into 3D printing. Traditional far-field electrospinning and electrospraying use a high voltage, over 10 kV, which causes issues such as high power consumption, severe electric discharge, electric shock, and fire safety hazard. On the contrary, the near-field method uses a less than 0.5 kV electrical power source to drive the EHD process. This provides the logic of integrating the unit into a 3D printer to form a new machine that overcomes the above-mentioned disadvantages associated with existing EHD machines.

As p-type semiconducting materials, Sb-Te alloys in a single-crystal state and in the conventional sintered polycrystalline form have been studied for TE energy conversion. Besides the TE application, Sb-Te-based compounds doped with Ge and Sn have been used as phase-change materials (PCMs) in rewritable optical data storage media [50]. Sb-Te materials can be processed by pulsed current sintering [51], electrodeposition [52], radio frequency magnetron sputtering [53], and thermal evaporation [54]. Sb-Te film can also be deposited from choline chloride–containing ionic liquids [55]. To make flexible energy converters, Sb-Te film was deposited on a Kapton™ polyimide substrate with preset curvatures followed by mechanical deformation [56] or by screen printing [57]. The Sb-Te alloy takes a distorted rock salt crystal structure [50, 58].

Through doping with some other elements, such as Ag, In, Bi, Se, Pb, and Cu [59–61], the TE, transport, and corrosion properties of Sb-Te alloys can be improved. The carrier concentration in Bi-containing Sb-Te alloys can be controlled by a tellurium-evaporation-annealing process [62]. Removal of oxygen through hydrogen reduction [63] and hot pressing [64] has been found to enhance the electrical conductive behavior of Sb-Te materials, while the thermal conductivity of Sb-Te alloys can be suppressed by controlling the crystal orientation [65] and grain size of the alloys [66].

Nanostructured Sb-Te alloys in the form of electrodeposited or sputtered thin films [67–70], particles [71], and nanowires (NWs) [72] show interesting properties. Nanocomposites consisting of lead telluride nanoinclusions in a bismuth antimony telluride matrix were made by an incipient wetness impregnation approach followed by hot pressing [73]. The study reveals that nanosized Pb-Te in the bulk Bi-Sb-Te matrix results in a special doping effect and changes the transport properties of the matrix alloy. Due to the interface effect, such a composite material system provides decreased thermal conductivity without sacrificing the electrical conductive property. Since the introduction of the low-temperature processing method for complex Sb-Te alloys [74], it is possible to make carbon/Sb-Te composites and polymer/Sb-Te composites. For example, through aerosol deposition at room temperature, Bi-Sb-Te can be deposited on the polyethylene terephthalate polymer to form TE legs [75]. Although vacuum infiltration followed by cold pressing was used for preparing multiwalled carbon nanotube (CNT)/Bi-Te composites [76], there is no early work reported on making partially carbonized nanofiber/Sb-Te composite materials.

Bismuth-based metallic alloys possess a semiconducting behavior. For example, bismuth telluride has a narrow energy band of E_g = 0.19 eV [77]. Another important bismuth alloy, Bi_2Se_3, shows a bandgap of around 0.35 eV [78]. They are suitable for TE energy conversion. They also find application in infrared (IR) radiation monitoring and temperature sensing. Bismuth telluride may be embedded in an insulating material, for example, alumina, to form a composite material. Such a composite offers enhanced TE figure of merit [79, 80]. It is expected that the electrical conductivity and Seebeck coefficient of the composites could be dramatically increased while the thermal conductivity could be kept as low as possible by selecting a suitable insulating matrix and controlling the bismuth alloy filler loading amount. This approach could provide a way to improve the TE performance of polymeric materials. Polymeric composite materials (ultrahigh-molecular-weight polyethylene based) with separated networks of CNT/Bi_2Te_3 hybrids were made [81]. However, the Seebeck coefficient of the polyethylene composite is only about 29 μV/K, which is much lower than that of the pure Bi-based TE materials. Simultaneous electrodeposition of polyaniline (PANI) and Bi_2Te_3 was conducted. It is found that the TE power of the composite material in the high temperature range (380 K ≈ 420 K) is higher than that of the pure PANI [82].

To increase the phonon scattering, incorporating nanoscale entities into the Bi-Te alloy matrix has also been considered. The dispersion of SiC nanoparticles in Bi_2Te_3 reduced the thermal conductivity but did not significantly increase the electrical resistance of the alloy, resulting in an 18% increase in the figure of merit [83]. CNT addition also resulted in increasing phonon scattering in Bi_2Te_3 [84]. Scattering of phonons is also anticipated in quantum-confined structures, as can be made by epitaxial growth [85], hydrothermal synthesis [86, 87], electrodeposition [88], and combined solution chemical method and thermal processing [89]. The mechanism for increased phonon scattering may be due to the interface reflection effect as discussed in Ref. [90] for Al_2O_3/Bi_2Te_3, and in Ref. [91] for silicon films.

Progress has been made in manufacturing Bi-based alloys and Bi_2Te_3-containing composite materials. Bi-based alloys are typically manufactured through spark plasma sintering and corrosion [92, 93], mechanical alloying [76, 94], organic-assisted growth [95], and interface reaction [96]. For composite materials, several major methods have been used, including the chemical reaction approach [97], mechanical alloying [98], plasma sintering [99], electrochemical synthesis [100, 101], and aerosol deposition [102]. Recently, carbon-based composite materials or composite materials of carbon mixed with bismuth-based alloys have caught much attention for TE applications [76, 84, 92, 103, 104]. Due to the low figure of merit, the TE performance of the composite materials still remains to be improved. Considering the promising development in this direction and some unsolved issues in this field, how to better integrate near-field EHD processing into 3D printing is worthy of discussion. In the following parts, how to manufacture composite materials containing nanofibers and nanoparticles by the innovative process in which EHD casting is integrated into 3D printing will be discussed first. Then, a scalable manufacturing process ensuring high process yield, process and product repeatability and reproducibility, and optimized quality control will be demonstrated. Finally, TE energy conversion and sensing functions of nanocomposite materials will be shown.

11.2 Manufacturing, Materials, and Characterization Method

The integration of EHD casting into 3D printing to make composite materials containing CNFs and bismuth telluride and/or antimony telluride particles is presented. Specifically, a new manufacturing machine is designed and made. This machine consists of an EHD casting unit installed on a 3D printer. Bismuth telluride and antimony telluride particle–loaded CNF composite materials are manufactured by the new machine. First, Bi_2Te_3 and Sb_2Te_3 particles were dispersed into a PAN polymer nanofiber mat through the EHD force–assisted 3D printing process. Then, stabilizing the PAN at intermediate temperatures and carbonizing the polymer through high-temperature annealing were carried out to obtain carbon fiber composite materials. The functions of the composite materials for energy conversions and sensing were validated. Fiber mats with controlled architectures or patterns can be manufactured through combined EHD casting and 3D printing process.

Characterization of the composite materials was conducted, which includes examining the structure of the materials and testing their energy conversion performances. Specifically, the scanning electron microscopic (SEM) analysis was performed to reveal the morphologies and compositions of the composite materials. Energy conversion properties of converting heat and photon energy into electricity were revealed. Hyperthermia behavior of converting electromagnetic wave energy into heat was also studied. The TE, hyperthermia, and photovoltaic effects of the composite materials were evaluated, with emphasis on the TE behavior. To make multiple component composites, an EHD unit consisting of a coaxial nozzle was designed. The EHD force–assisted 3D printing machine was made by attaching the EHD unit to a 3D printer. The as-mentioned composite materials were made using the new machine. The Seebeck coefficient of the composites was measured using an HP 34401A multimeter. The electrical conductivity maps of the composites were generated using an atomic force microscope (AFM). The scope of the manufacturing and characterization is given in the flowchart shown in Fig. 11.1.

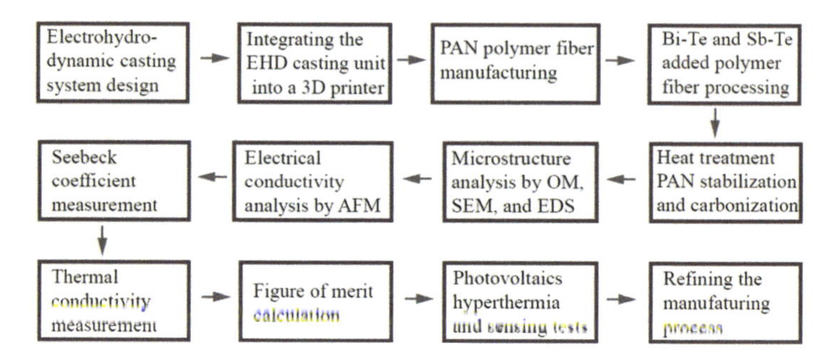

Figure 11.1 The manufacturing and characterization flowchart.

11.3 Integrating Near-Field Electrohydrodynamic Casting into 3D Printing

First of all, near-field EHD casting was performed to make sure the size and orientation of nanofibers can be controlled. Our preliminary work indicated that by changing the voltage and the distance of the fibers from the fiber collector, the diameters of the fibers can be easily controlled in the range of several tens of nanometers to several hundred nanometers, as shown in Fig. 11.2. Figure 11.2a shows the randomly distributed micro- and nanofibers prepared by far-field EHD casting at a high voltage, of 15 kV. In Fig. 11.2b, the result of controlled deposition of nanofibers by integrating near-field EHD processing into 3D printing is shown. The studies on integrating near-field EHD processing into 3D printing for making different polymer fibers generate meaningful results. The obtained polyvinylpyrrolidone (PVP) nanofibers with good alignment are shown in Fig. 11.2c.

More precise controlling of the size of the nanofibers was performed to understand the effect of manufacturing parameters on the quality of the fibers. The orientation of the nanofibers can be better controlled by optimizing the near-field EHD casting parameters. The regular fiber alignment is critical for some applications, such as guided tissue regeneration, fast responsive sensing, and high-efficiency energy conversions. Therefore, it is necessary to use a near-field EHD process to generate nanostructure. To obtain better controllability of the fiber orientation, the voltage was kept in the range of 0.5 to 1.2 kV. The near-field condition was maintained by keeping the space between the positively charged nozzle and the negatively charged fiber collector to as close as 500 microns, which is the gap between the tip of the extruder and the flatbed of the 3D printer.

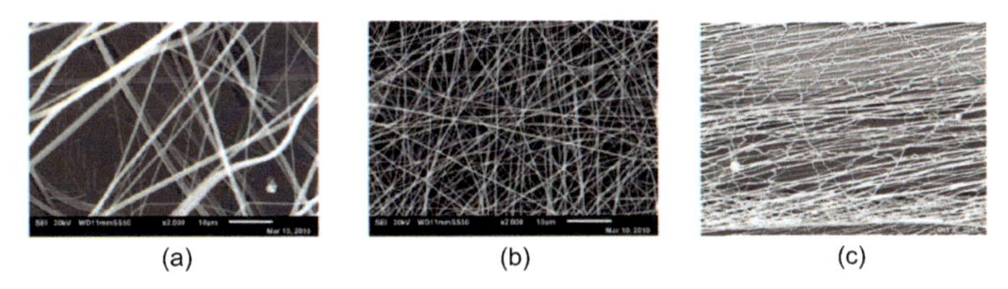

Figure 11.2 Micro- and nanofiber structures made by far-field EHD casting and integrating EHD casting into 3D printing: (a) polyacrylonitrile (PAN) microfibers and nanofibers prepared by far-field EHD casting, (b) controlled deposition of polyacrylonitrile nanofibers by integrating far-field EHD casting and 3D printing, and (c) controlled deposition of polyvinylpyrrolidone (PVP) nanofibers by integrating near-field EHD casting and 3D printing.

11.4 The New Manufacturing Machine Design

A coaxial nozzle was designed first for the EHD casting. Secondly, the near-field EHD casting unit was integrated into a 3D printer. The integrated machine allowed fiber materials to be placed much more precisely than those shown in Fig. 11.2a or 11.2b. It must be noted that if just 3D printing was used, the resolution would be only in the millimeter range. Consequently, no nanofibers could be made directly through 3D printing. Therefore, the design of a new manufacturing machine by attaching a self-made EHD casting unit to a 3D printer as shown in Fig. 11.3a was implemented. This machine was operated by using the 3D printer for generating preprogrammed *x-y-z* three-directional motions so that the nanofibers that came out from the EHD casting unit could be placed in the way as designed by the computer programming. The fluid control unit as schematically shown allowed polymer solutions to be delivered in the required amount.

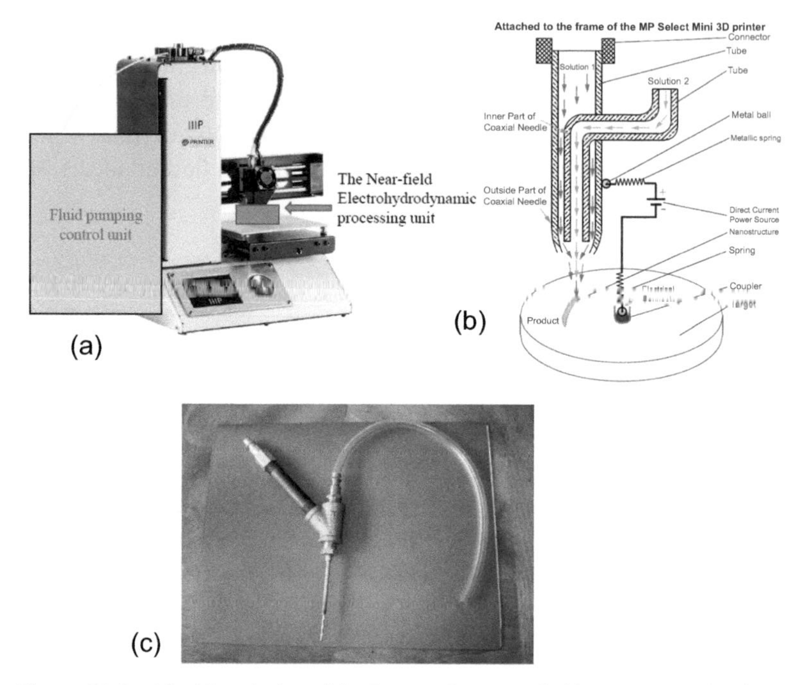

Figure 11.3 Machine designed for integrating near-field EHD processing into 3D printing: (a) the concept of an integrated machine, (b) schematic of a coaxial nozzle for multilayered nanofiber casting, and (c) a coaxial nozzle consisting of a brass base and stainless-steel jets.

To show that integrating EHD casting into 3D printing could lead to a new additive manufacturing technology that allows the production of aligned 3D nanostructures with multiple components, nozzles with a coaxial structure were made. Figure 11.3b schematically illustrates how different solutions can be delivered to the tip of the coaxial nozzle, and Fig. 11.3c shows the physical model of a typical coaxial nozzle for making multiple component nanocomposite materials in the machine.

11.5 Manufacturing Nanoparticle-Loaded Composite Fibers

The solvent, DMF, was purchased from Alfa Aesar. Bi_2Te_3 and Sb_2Te_3 raw powders were purchased from Sigma Aldrich. The PAN polymer was obtained from Scientific Polymer, Inc., Ontario, New York. The PAN solution was prepared by adding approximately 10% in weight of the PAN powder to the DMF solvent. For a typical experiment, 10 g of PAN polymer was added to 90 mL of DMF solvent and stirred for an hour to allow the PAN powder to dissolve in the solvent. Then 10 g of refined Bi_2Te_3 or Sb_2Te_3 powders with the nominal size of less than 1 μm were added to the PAN solution.

The solution containing the PAN polymer, Bi_2Te_3 or Sb_2Te_3 powders, and DMF solvent was filled in a plastic vessel as the sheath fluid. The core fluid consisting only of the PAN polymer in DMF was also loaded into another plastic vessel. The two plastic vessels were connected with two precision syringe pumps made by Chemyx, Inc., Stafford, Texas, USA. The two pumps can precisely control the injection flow rate of the two solutions. The distance between the tip of the nozzle for injection as shown in Fig. 11.3c and the receiving target was kept at 0.5 mm. A constant flow rate of the solutions injected by the syringe pumps was kept at 0.1 mL/min. Under ambient temperature, pressure, and humidity conditions, a DC voltage of 0.8 kV was applied at the tip of the nozzle to electrify the solutions. This electrified state allows the mixture material to overcome the surface tension of the solutions. The cast fibers were collected on the flatbed of the 3D printer as shown in Fig. 11.3a. The electric potential difference between the tip of the nozzle and the ground collector led the charged jet to cast fibers on the collector continuously. The 3D printer allowed the control of the orientation of the cast nanofibers. In other words, the collector was set on an *x-y-z* table, which can translate into *x*-, *y*-, and *z*-directions. From this processing step, the product collected was a composite mat containing Bi-Te or Sb-Te nanoparticles within the PAN nanofibers.

11.6 Stabilization and Carbonization Heat Treatment

To form an electrically conductive network around the Bi-Te and Sb-Te particles, the PAN nanofibers were stabilized and converted into partially carbonized nanofibers through heat treatment using the facility shown on the left in Fig. 11.4.

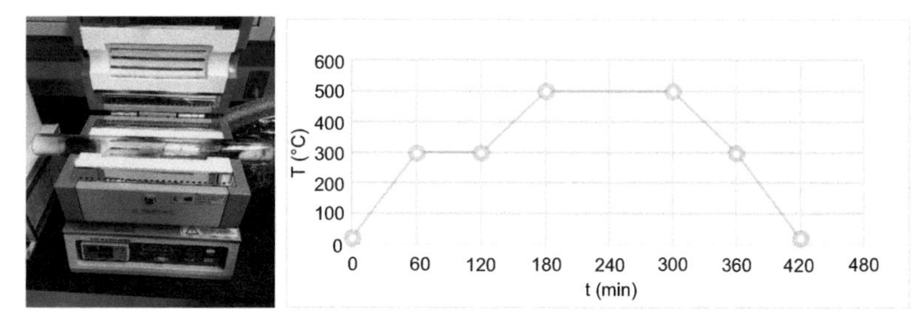

Figure 11.4 The heat treatment instrument setup (left) and the temperature profile (right).

The heat treatment can be divided into two steps. First, the composite specimen was put in the quartz tube and heated up to 300°C in an ambient atmosphere. After being kept at 300°C for 1 h, the composite fiber material was heated up slowly to 500°C. Hydrogen was inducted carefully into the chamber during the heating from 300°C to 500°C. This prevents Bi-Te or Sb-Te from being oxidized. The PAN fiber started carbonizing partially. After the material was heat-treated at 500°C for 1 h, it was cooled down naturally with the furnace to room temperature. It is believed that the PAN polymer underwent cyclization and oxidation when it was heated in the temperature range from 200°C to 300°C in air [105–107]. In this work, a stabilizing temperature of 300°C was used. The PAN molecules were cyclized and transformed into a nonmeltable ladder structure as reported earlier [108–110]. Some oxygen in the functional groups such as =O and –OH attached to the backbone structure of cyclized polymeric carbon [111, 112] was removed due to exposure to the hydrogen atmosphere. To determine the elemental compositions of the composite materials, energy-dispersive X-ray diffraction spectra (EDS) were obtained by mapping the selected areas on the materials.

11.7 Thermoelectric Property and Sensing Behavior Characterization

The TE response of the Bi_2Te_3/CNF and Sb_2Te_3/CNF composites was examined using a thermal wave testing method. Here we focus on the results obtained from the tests on the Bi_2Te_3/CNF composite. As shown in Fig. 11.5a, a 1.5 kW Drill Master heating gun was used to blow hot air toward the composite material specimen. A preliminary study of a Bi_2Te_3/CNF specimen with the width × length dimension of 15 mm × 50 mm shows promising results. When the 110 V AC power supply to the heat gun was on, the open circuit voltage of the specimen dropped by about 0.5 V, as shown in Fig. 11.5b. The temperature of the hot air that reached the surface of the specimen was around 150°C. The room temperature of the cold end was about 23°C under the test condition. As known, the Seebeck coefficient can be calculated.

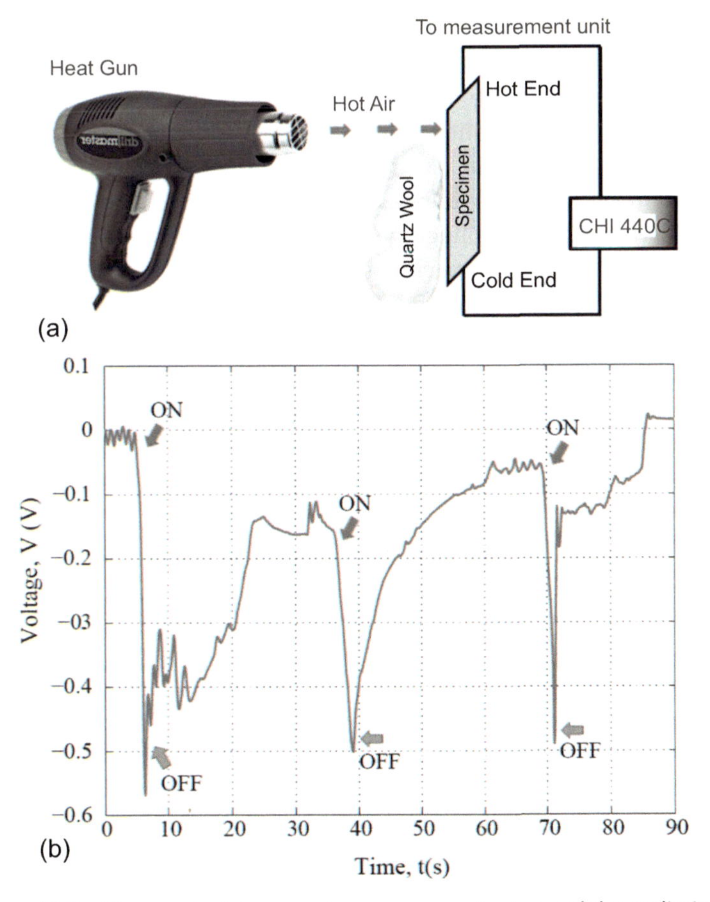

Figure 11.5 Thermoelectric property measurement setup and the preliminary study results: (a) thermal wave test setup and (b) thermoelectric response of the Bi_2Te_3/CNF composite material. Republished with permission of ASME, from Ref. [113] (2018); permission conveyed through Copyright Clearance Center, Inc.

The estimated value of S for the Bi_2Te_3/CNF material is about −3.937 mV/K. This result provides us important information. First, the composite material shows a general n-type behavior because the voltage dropped to a negative value when the specimen was heated. Second, the best carbon-based composite materials—the stacked graphene sheets–based TE material—can only reach a Seebeck coefficient value of −90 μV/K [104]. The results from our current study show that the Bi_2Te_3/CNF composite mat is more than 40 times stronger than the currently existing carbon composite materials.

On the basis of the strong TE effect of the Bi_2Te_3/CNF composite material, tests on the sample as a prototype sensor to monitor the warm air flow due to exhaling and inhaling of humans were made. The results from the preliminary study are shown in Fig. 11.6. First, a relatively short exhaling phase (5 s) was followed by a longer inhaling stage (15 s) in a complete cycle, and the test results are plotted in Fig. 11.6a. Obviously, the warm air from the exhaling stage caused the negative voltage generation. The surface temperature of the hot end of the specimen can be monitored by an INF165 IR thermometer available from UEi Test Instruments, Beaverton, Oregon, USA. The hot-end surface temperature of the specimen was around 31°C. And the cold-end surface temperature was about 26°C. From the voltage difference in the exhaling and inhaling stages, it was found that the Seebeck coefficient was approximately 4 mV/K. This value is consistent with the result as presented in Fig. 11.5b. The test results related to a short exhaling phase (5 s) followed by a short inhaling stage (5 s) in a full cycle are plotted in Fig. 11.6b. It was found that due to the short recovery time for the temperature in the inhaling cycle, the voltage changed less than that shown in Fig. 11.6a. The data revealed a general trend of decreasing in the open circuit voltage.

The breathing pattern of a patient with a coughing symptom may be monitored. The results are shown in Fig. 11.6c. Evidently, the breathe-in signals were momentarily strong because the air came out due to the coughing, as shown by the sharp drop in the voltage. Because of the reflexing actions, the period right after the breathe-in and the following breath-out signals reveal irregularity, as evident by the zig-zag patterns marked by the green circles in Fig. 11.6c. The implication of the test results in Fig. 11.6 is that an inexpensive sensor may be built based on the highly sensitive TE responses of the Bi_2Te_3/CNF composite material.

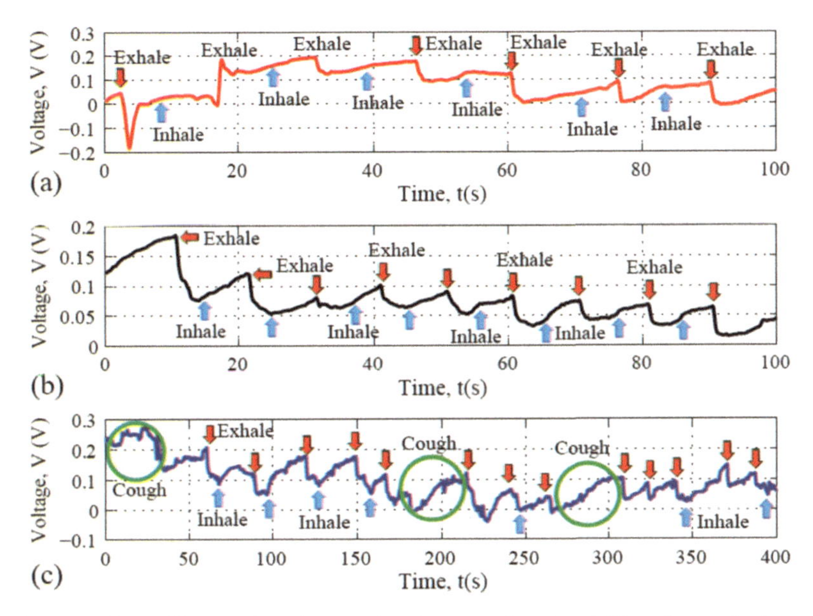

Figure 11.6 Thermoelectric results showing the feasibility of monitoring breathe in and breathe out: (a) short exhaling followed by long inhaling, (b) exhaling and inhaling in equal time, and (c) monitoring the coughing breathe-in and breathe-out patterns. Republished with permission of ASME, from Ref. [113] (2018); permission conveyed through Copyright Clearance Center, Inc.

11.8 Scalable Manufacturing of Composite Material Mats

It is meaningful to explore high-speed EHD casting combined with 3D printing toward a scalable process. On the basis of the facility setup as shown in Fig. 11.2a, the function of the receiving plate attached to an *x*-*y*-*z* table and the table was controlled to translate in three directions. The translational motions allowed the nanofiber to be cast onto the target quicker and dry faster. A centimeter-level thickness composite mat was obtained by this system, and the images are shown in Fig. 11.7. Figure 11.7a provides a global view of how the thick mat is like. Figure 11.7b, a slightly higher magnification (200X) image, shows the crease pattern due to the fast deposition of the nanofibers. At 500X, the uniform distribution of particles in the nanofiber mat can be revealed. The SEM observation at 1000X illustrates the connection of the microparticles and nanofibers. From the preliminary studies, the process seems scalable. More comprehensive studies were performed to understand the effects of fluid pressure, 3D printing speed, and the EHD parameters on the scalability of the manufacturing.

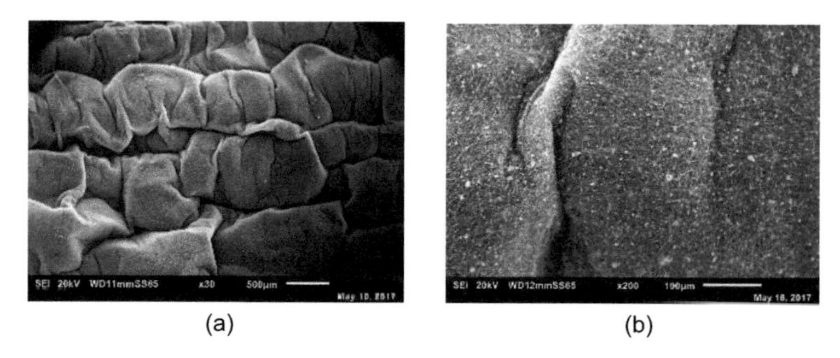

Figure 11.7 SEM images of carbonized composite nanofiber mats: (a) at 30X and (b) at 200X.

11.9 Conclusions

Integrating EHD casting into 3D printing represents a new manufacturing technology. This new technology has been successfully used for making Bi-Te/PAN nanofiber composite materials. It is also scalable to produce fiber mats. The high-temperature heat treatment converts the composite nanofibers to polymeric CNFs. SEM analysis reveals the uniform distribution of the Bi-Te particles in the partially carbonized nanofiber matrix. The CNFs form networks connecting the Bi-Te particles. TE response tests indicate that the absolute value of the Seebeck coefficient of the Bi-Te/CNF, 4 mV/K, is much higher than that of the currently available carbon-based TE materials or Bi-Te alloys.

Photovoltaic response measurement of the heat-treated Bi-Te/CNF composite mat specimen shows a semiconducting behavior. Only the CNF is considered to respond to the visible light because of the polymeric carbon network with a π-conjugated structure. Since Bi-Te is IR sensitive, it has the potential to be used as photoelectric sensors in the wide spectrum covering the frequency range from IR to visible light.

References

1. Taylor, G. (1969). Electrically driven jets, *Proc. R. Soc. A, Math. Phys. Sci.*, **313**, pp. 453–475.
2. Melcher, J. R. (1963). *Field-Couple Surface Waves: A Comparative Study of Surface Coupled Electrohydrodynamic and Magnetohydrodynamic Systems*, Cambridge, Massachusetts: The MIT Press, pp. 1–63.
3. VillaVelázquez-Mendoza, C. I., Mendoza-Barraza, S. S., Rodriguez, J. L., Levy-Padilla, M. I., Ibarra-Galván, V., Zamudio-Ojeda, A. (2016). Simultaneous synthesis of β-Si_3N_4 nanofibers and pea-pods and hand-fan like Si_2N_2O nanostructures by the CVD method, *Mater. Lett.*, **175**, pp. 139–142.
4. Maldonado, J. R., Peckerar, M. (2016). X-Ray lithography: some history, current status and future prospects, *Microelectron. Eng.*, **161**, pp. 87–93.
5. Stoychev, G. V., Okhrimenko, D. V., Appelhans, D., Voit, B. (2016). Electron beam-induced formation of crystalline nanoparticle chains from amorphous cadmium hydroxide nanofibers, *J. Colloid Interface Sci.*, **461**, pp. 122–127.
6. Subbiah, T., Bhat, G. S., Tock, R. W., Parameswaran, S. (2005). Electrospinning of nanofibers, *J. Appl. Polym. Sci.*, **96**, pp. 557–559.
7. Gan, Y. X., Chen, A. D., Gan, R. N., Hamdan, A. S. (2017). Energy conversion behaviors of antimony telluride particle loaded partially carbonized nanofiber composite mat manufactured by electrohydrodynamic casting, *Microelectron. Eng.*, **181**, pp. 16–21.
8. Gan, Y. X. (2018). A review of electrohydrodynamic casting energy conversion polymer composites, *AIMS Mater. Sci.*, **5**, pp. 206–225.
9. Han, Y., Wei, C., Dong, J. (2015). Droplet formation and settlement of phase-change ink in high resolution electrohydrodynamic (EHD) 3D printing, *J. Manuf. Proc.*, **20**, pp. 485–491.
10. Han, Y., Dong, J. (2017). High-resolution electrohydrodynamic (EHD) direct printing of molten metal, *Procedia Manuf.*, **10**, pp. 845–850.
11. Zhang, Y., Huang, Z. M., Xu, X., Lim, C. T., Ramakrishna, S. (2004). Preparation of core-shell structured PCL-r-gelatin bi-component

nanofibers by coaxial electrospinning, *Chem. Mater.*, **16**(18), pp. 3406–3409.

12. Loscertales, I. G., Barrero, A., Guerrero, I., Cortijo, R., Marquez, M., Ganan-Calvo, A. M. (2002). Micro/nano encapsulation via electrified coaxial liquid jets, *Science*, **295**(5560), pp. 1695–1698.
13. Kurban, Z., Lovell, A., Bennington, S. M., Jenkins, D. W. K., Ryan, K. R., Jones, M. O., Skipper, N. T., David, W. I. F. (2010). A solution selection model for coaxial electrospinning and its application to nanostructured hydrogen storage materials, *J. Phys. Chem. C*, **114**(49), pp. 21201–21213.
14. Wang, C., Yan, K. W., Lin, Y. D., Hsieh, P. C. H. (2010). Biodegradable core/shell fibers by coaxial electrospinning: processing, fiber characterization, and its application in sustained drug release, *Macromolecules*, **43**(15), pp. 6389–6397.
15. Zhang, Y. Z., Wang, X., Feng, Y., Li, J., Lim, C. T., Ramakrishna, S. (2006). Coaxial electrospinning of (fluorescein isothiocyanate-conjugated bovine serum albumin)-encapsulated poly(ε-caprolactone) nanofibers for sustained release, *Biomacromolecules*, **7**, pp. 1049–1057.
16. Zhang, H., Zhao, C. G., Zhao, Y. H., Tang, G. W., Yuan, X. Y. (2010). Electrospinning of ultrafine core/shell fibers for biomedical applications, *Sci. China Chem.*, **53**(6), pp. 1246–1254.
17. Li, F., Zhao, Y., Song, Y. (2010). Core-shell nanofibers: nano channel and capsule by coaxial electrospinning, in *Nanofibers* (A. Kumar ed.), Croatia: InTech, pp. 419–438.
18. Chan, K. H. K., Kotaki, M. (2009). Fabrication and morphology control of poly(methyl methacrylate) hollow structures via coaxial electrospinning, *J. Appl. Polym. Sci.*, **111**, pp. 408–416.
19. Chen, H., Wang, N., Di, J., Zhao, Y., Song, Y., Jiang, L. (2010). Nanowire-in-microtube structured core/shell fibers via multifluidic coaxial electrospinning, *Langmuir*, **26**(13), pp. 11291–11296.
20. Yu, J. H., Fridrikh, S. V., Rutledge, G. C. (2004). Production of submicron diameter fibers by two-fluids electrospinning, *Adv. Mater.*, **16**(17), pp. 1562–1566.
21. Sun, B., Long, Y. Z., Zhang, H. D., Li, M. M., Duvail, J. L., Jiang, X. Y., Yin, H. L. (2014). Advances in three-dimensional nanofibrous macrostructures via electrospinning, *Prog. Polym. Sci.*, **39**, pp. 862–890.
22. Zhang, Y., Tse, C., Rouholamin, D., Smith, P. J. (2012). Scaffolds for tissue engineering produced by inkjet printing, *Cent. Eur. J. Eng.*, **2**, pp. 325–335.

23. Park, T. H., Shuler, M. L. (2003). Integration of cell culture and microfabrication technology, *Biotechnol. Prog.*, **19**, pp. 243–253.

24. Lee, M., Kim, H. Y. (2014). Toward nanoscale three-dimensional printing: nanowalls built of electrospun nanofibers, *Langmuir*, **30**, pp. 1210–1214.

25. Mandrycky, C., Wang, Z., Kim, K., Kim, D. H. (2016). 3D bioprinting for engineering complex tissues, *Biotechnol. Adv.*, **34**(4), pp. 422–434.

26. Huang, C., Jian, G., DeLisio, J. B., Wang, H., Zachariah, M. R. (2015). Electrospray deposition of energetic polymer nanocomposites with high mass particle loadings: a prelude to 3D printing of rocket motors, *Adv. Eng. Mater.*, **17**, pp. 95–101.

27. Liu, Y., Pollaor, S., Wu, Y. (2015). Electrohydrodynamic processing of p-type transparent conducting oxides, *J. Nanomater.*, **2015**, Article ID 423157.

28. Sun, J., Zhou, W., Huang, D., Fuh, J. Y. H., Hong, G. S. (2015). An overview of 3D printing technologies for food fabrication, *Food Bioprocess Technol.*, **8**, pp. 1605–1615.

29. Mironov, V., Trusk, T., Kasyanov, V., Little, S., Swaja, R., Markwald, R. (2009). Biofabrication: a 21st century manufacturing paradigm, *Biofabrication*, **1**, 022001 (16pp).

30. Visser, J., Peters, B., Burger, T. J., Boomstra, J., Dhert, W. J. A., Melchels, F. P. W., Malda, J. (2013). Biofabrication of multi-material anatomically shaped tissue constructs, *Biofabrication*, **5**, 035007 (9pp).

31. Mittal, A., Negi, P., Garkhal, K., Verma, S., Kumar, N. (2010). Integration of porosity and bio-functionalization to form a 3D scaffold: cell culture studies and *in vitro* degradation, *Biomed. Mater.*, **5**, pp. 1–16.

32. Ozbolat, I., Yu, Y. (2013). Bioprinting towards organ fabrication: challenges and future trends, *IEEE Trans. Biomed. Eng.*, **60**(3), pp. 691–699.

33. Mironov, V., Rels, N., Derby, B. (2006). Bioprinting: a beginning, *Tissue Eng.*, **12**, pp. 631–634.

34. Catros, S., Guillemot, F., Nandakumar, A., Ziane, S., Moroni, L., Habibovic, P., Blitterswijk, C. V., Rousseau, B., Chassande, O., Amedee, J., Fricain, J. C. (2011). Layer-by-layer tissue microfabrication supports cell proliferation in vitro and in vivo, *Tissue Eng.*, **18**, pp. 1–9.

35. Vozzi, G., Tirella, A., Ahluwalia, A. (2012). Rapid prototyping composite and complex scaffolds with PAM2, *Methods Mol. Biol.*, **868**, pp. 59–70.

36. Shim, J. H., Yoon, M. C., Jeong, C. M., Jang, J., Jeong, S. I., Cho, D. W., Huh, J. B. (2014). Efficacy of rhBMP-2 loaded PCL/PLGA/β-TCP guided

bone regeneration membrane fabricated by 3D printing technology for reconstruction of calvaria defects in rabbit, *Biomed. Mater.*, **9**, p. 065006.

37. Laudenslager, M. J., Sigmund, W. M. (2011). Developments in electrohydrodynamic forming: fabricating nanomaterials from charged liquids via electrospinning and electrospraying, *Am. Ceram. Soc. Bull.*, **90**, pp. 23–27.
38. Martins, A., Chung, S., Pedro, A. J., Sousa, R. A., Marques, A. P., Reis, R. L., Neves, N. M. (2009). Hierarchical starch-based fibrous scaffold for bone tissue engineering applications, *J. Tissue Eng. Regener. Med.*, **3**, pp. 37–42.
39. Zhu, W., Masood, F., O'Brien, J., Zhang, L. G. (2015). Highly aligned nanocomposite scaffolds by electrospinning and electrospraying for neural tissue regeneration, *Nanomed. Nanotechnol. Biol. Med.*, **11**, pp. 693–704.
40. Erisken, C., Kalyon, D. M., Wang, H. (2008). Functionally graded electrospun polycaprolactone and b-tricalcium phosphate nanocomposites for tissue engineering applications, *Biomaterials*, **29**, pp. 4065–4073.
41. Giannitelli, S. M., Mozetic, P., Trombetta, M., Rainer, A. (2015). Combined additive manufacturing approaches in tissue engineering, *Acta Biomater.*, **24**, pp. 1–11.
42. Xu, T., Binder, K. W., Albanna, M. Z., Dice, D., Zhao, W., Yoo, J. J., Atala, A. (2013). Hybrid printing of mechanically and biologically improved constructs for cartilage tissue engineering applications, *Biofabrication*, **5**, 015001 (10pp).
43. Nam, J., Huang, Y., Agarwal, S., Lannutti, J. (2007). Improved cellular infiltration in electrospun fiber via engineered porosity, *Tissue Eng.*, **13**, pp. 2249–2257.
44. Abdelaal, O. A. M., Darwish, S. M. H. (2013). Review of rapid prototyping techniques for tissue engineering scaffolds fabrication, in *Characterization and Development of Biosystems and Biomaterials, Advanced Structured Materials*, Vol. 29, Springer-Verlag Berlin Heidelberg, pp. 33–54.
45. Zhu, W., O'Brien, C., O'Brien, J. R., Zhang, L. G. (2014). 3D nano/microfabrication techniques and nanobiomaterials for neural tissue regeneration, *Nanomedicine*, **9**, pp. 859–875.
46. Karande, T. S., Ong, J. L., Agrawal, C. M. (2004). Diffusion in musculoskeletal tissue engineering scaffolds: design issues related to

porosity, permeability, architecture, and nutrient mixing, *Ann. Biomed. Eng.*, **32** pp. 1728–1743.

47. Jung, J. W., Lee, H., Hong, J. M., Park, J. H., Shim, J. H., Choi, T. H., Cho, D. W. (2015). A new method of fabricating a blend scaffold using an indirect three dimensional printing technique, *Biofabrication*, **7**, p. 045003.

48. Kim, J. T., Seol, S. K., Pyo, J., Lee, J. S., Je, J. H., Margaritondo, G. (2011). Three-dimensional writing of conducting polymer nanowire arrays by meniscus-guided polymerization, *Adv. Mater.*, **23**, pp. 1968–1970.

49. Pham, Q. P., Sharma, U., Mikos, A. G. (2006). Electrospinning of polymeric nanofibers for tissue engineering applications: a review, *Tissue Eng.*, **12**, pp. 2249–2257.

50. Rosenthal, T., Welzmiller, S., Neudert, L., Urban, F., Fitch, A., Oeckler, O. (2014). Novel superstructure of the rock salt type and element distribution in germanium tin antimony tellurides, *J. Solid State Chem.*, **219**, pp. 108–117.

51. Kitagawa, H., Takimura, K., Ido, S., Morito, S., Kikuchi, K. (2017). Thermoelectric properties of crystal-aligned bismuth antimony tellurides prepared by pulse-current sintering under cyclic uniaxial pressure, *J. Alloys Compd.*, **692**, pp. 388–394.

52. Hatsuta, N., Takemori, D., Takashiri, M. (2016). Effect of thermal annealing on the structural and thermoelectric properties of electrodeposited antimony telluride thin films, *J. Alloys Compd.*, **685**, pp. 147–152.

53. Sasaki, Y., Takashiri, M. (2016). Effects of Cr interlayer thickness on adhesive, structural, and thermoelectric properties of antimony telluride thin films deposited by radio-frequency magnetron sputtering, *Thin Solid Films*, **619**, pp. 195–201.

54. Takashiri, M., Hamada, J. (2016). Bismuth antimony telluride thin films with unique crystal orientation by two-step method, *J. Alloys Compd.*, **683**, pp. 276–281.

55. Catrangiu, A. S., Sin, I., Prioteasa, P., Cotarta, A., Cojocaru, A., Anicai, L., Visan, T. (2016). Studies of antimony telluride and copper telluride films electrodeposition from choline chloride containing ionic liquids, *Thin Solid Films*, **611**, pp. 88–100.

56. Masayuki, K., Takashiri, M. (2015). Investigation of the effects of compressive and tensile strain on n-type bismuth telluride and p-type antimony telluride nanocrystalline thin films for use in flexible thermoelectric generators, *J. Alloys Compd.*, **653**, pp. 480–485.

57. Catlin, G. C., Tripathi, R., Nunes, G., Lynch, P. B., Jones, H. D., Schmitt, D. C. (2017). An additive approach to low temperature zero pressure sintering of bismuth antimony telluride thermoelectric materials, *J. Power Sources*, **343**, pp. 316–321.

58. Urban, P., Schneider, M. N., Oeckler, O. (2015). Temperature dependent ordering phenomena in single crystals of germanium antimony tellurides, *J. Solid State Chem.*, **227**, pp. 223–231.

59. Hu, L. P., Zhu, T. J., Yue, X. Q., Liu, X. H., Wang, Y. G., Xu, Z. J., Zhao, X. B. (2015). Enhanced figure of merit in antimony telluride thermoelectric materials by In–Ag Co-alloying for mid-temperature power generation, *Acta Mater.*, **85**, pp. 270–278.

60. Lee, W. Y., Park, N. W., Hong, J. E., Yoon, S. G., Koh, J. H., Lee, S. K. (2015). Effect of electronic contribution on temperature-dependent thermal transport of antimony telluride thin film, *J. Alloys Compd.*, **620**, pp. 120–124.

61. Rosalbino, F., Carlini, R., Zanicchi, G., Scavino, G. (2013). Microstructural characterization and corrosion behavior of lead, bismuth and antimony tellurides prepared by melting, *J. Alloys Compd.*, **567**, pp. 26–32.

62. Kim, D. H., Kwon, I. H., Kim, C., Han, B., Im, H. J., Kim, H. (2013). Tellurium-evaporation-annealing for p-type bismuth-antimony-telluride thermoelectric materials, *J. Alloys Compd.*, **548**, pp. 126–132.

63. Bochentyn, B., Miruszewski, T., Karczewski, J., Kusz, B. (2016). Thermoelectric properties of bismuth-antimony-telluride alloys obtained by reduction of oxide reagents, *Mater. Chem. Phys.*, **177**, pp. 353–359.

64. Qiu, W., Yang, S., Zhao, X. (2011). Effect of hot-press treatment on electrochemically deposited antimony telluride film, *Thin Solid Films*, **519**, pp. 6399–6402.

65. Takashiri, M., Tanaka, S., Miyazaki, K. (2010). Improved thermoelectric performance of highly-oriented nanocrystalline bismuth antimony telluride thin films, *Thin Solid Films*, **519**, pp. 619–624.

66. Takashiri, M., Tanaka, S., Hagino, H., Miyazaki, K. (2014). Strain and grain size effects on thermal transport in highly-oriented nanocrystalline bismuth antimony telluride thin films, *Int. J. Heat Mass Transfer*, **76**, pp. 376–384.

67. Lim, S. K., Kim, M. Y., Oh, T. S. (2009). Thermoelectric properties of the bismuth-antimony-telluride and the antimony-telluride films processed by electrodeposition for micro-device applications, *Thin Solid Films*, **517**, pp. 4199–4203.

68. Jung, H., Myung, N. V. (2011). Electrodeposition of antimony telluride thin films from acidic nitrate-tartrate baths, *Electrochim. Acta*, **56**, pp. 5611–5615.

69. Fan, P., Chen, T., Zheng, Z., Zhang, D., Cai, X., Cai, Z., Huang, Y. (2013). The influence of Bi doping in the thermoelectric properties of Co-sputtering deposited bismuth antimony telluride thin films, *Mater. Res. Bull.*, **48**, pp. 333–336.

70. Lensch-Falk, J. L., Banga, D., Hopkins, P. E., Robinson, D. B., Stavila, V., Sharma, P. A., Medlin, D. L. (2012). Electrodeposition and characterization of nano-crystalline antimony telluride thin films, *Thin Solid Films*, **520**, pp. 6109–6117.

71. Takashiri, M., Tanaka, S., Miyazaki, K. (2013). Growth of single-crystalline bismuth antimony telluride nanoplates on the surface of nanoparticle thin films, *J. Cryst. Growth*, **372**, pp. 199–204.

72. Kim, B. G., Choi, S. M., Lee, M. H., Seo, W. S., Lee, H. L., Hyun, S. H., Jeong, S. M. (2015). Facile fabrication of silicon and aluminum oxide nanotubes using antimony telluride nanowires as templates, *Ceram. Int.*, **41**, pp. 12246–12252.

73. Ganguly, S., Zhou, C., Morelli, D., Sakamoto, J., Uher, C., Brock, S. L. (2011). Synthesis and evaluation of lead telluride/bismuth antimony telluride nanocomposites for thermoelectric applications, *J. Solid State Chem.*, **184**, pp. 3195–3201.

74. Li, J., Chen, Z., Wang, X., Proserpio, D. M. (1997). A novel two-dimensional mercury antimony telluride: low temperature synthesis and characterization of $RbHgSbTe_3$, *J. Alloys Compd.*, **262–263**, pp. 28–33.

75. Baba, S., Sato, H., Huang, L., Uritani, A., Funahashi, R., Akedo, J. (2014). Formation and characterization of polyethylene terephthalate-based $(Bi_{0.15}Sb_{0.85})_2Te_3$ thermoelectric modules with $CoSb_3$ adhesion layer by aerosol deposition, *J. Alloys Compd.*, **589**, pp. 56–60.

76. Bark, H., Kim, J. S., Kim, H., Yim, J. H., Lee, H. (2013). Effect of multiwalled carbon nanotubes on the thermoelectric properties of a bismuth telluride matrix, *Curr. Appl. Phys.*, **13**, pp. S111–S114.

77. Zhang, H. T., Luo, X. G., Wang, C. H., Xiong, Y. M., Li, S. Y., Chen, X. H. (2004). Characterization of nanocrystalline bismuth telluride (Bi_2Te_3) synthesized by a hydrothermal method, *J. Cryst. Growth*, **265**, pp. 558–562.

78. Sun, Y., Cheng, H., Gao, S., Liu, Q., Sun, Z., Xiao, C., Wu, C., Wei, S., Xie, Y. (2012). Atomically thick bismuth selenide freestanding single layers

achieving enhanced thermoelectric energy harvesting, *J. Am. Chem. Soc.*, **134**, pp. 20294–20297.

79. Prieto, A. L., Sander, M. S., Martin-Gonzalez, M. S., Gronsky, R., Sands, T., Stacy, A. M. (2001). Electrodeposition of ordered Bi_2Te_3 nanowire arrays, *J. Am. Chem. Soc.*, **123**(29), pp. 7160–7161.

80. Borca-Tasciuc, D. A., Chen, G., Prieto, A., Martín-González, M. S., Stacy, A., Sands, T., Ryan, M. A., Fleurial, J. P. (2004). Thermal properties of electrodeposited bismuth telluride nanowires embedded in amorphous alumina, *Appl. Phys. Lett.*, **85**(24), pp. 6001–6003.

81. Pang, H., Piao, Y. Y., Tan, Y. Q., Jiang, G. Y., Wang, J. H., Li, Z. M. (2013). Thermoelectric behavior of segregated conductive polymer composites with hybrid fillers of carbon nanotube and bismuth telluride, *Mater. Lett.*, **107**, pp. 150–153.

82. Chatterjee, K., Suresh, A., Ganguly, S., Kargupta, K., Banerjee, D. (2009). Synthesis and characterization of an electro-deposited polyaniline-bismuth telluride nanocomposite: a novel thermoelectric material, *Mater. Charact.*, **60**, pp. 1597–1601.

83. Li, J. F., Liu, J. (2006). Effect of nano-SiC dispersion on thermoelectric properties of Bi_2Te_3 polycrystals, *Phys. Status Solidi A*, **203**(15), pp. 3768–3773.

84. Kim, K. T., Choi, S. Y., Shin, E. H., Moon, K. S., Koo, H. Y., Lee, G. G., Ha, G. H. (2013). The influence of CNTs on the thermoelectric properties of a CNT/Bi_2Te_3 composite, *Carbon*, **52**, pp. 541–549.

85. Lu, W., Ding, Y., Chen, Y., Wang, Z. L., Fang, J. (2005). Bismuth telluride hexagonal nanoplatelets and their two-step epitaxial growth, *J. Am. Chem. Soc.*, **127**(28), pp. 10112–10116.

86. Sumithra, S., Takas, N. J., Misra, D. K., Nolting, W. M., Poudeu, P. F. P., Stokes, K. L. (2011). Enhancement in thermoelectric figure of merit in nanostructured Bi_2Te_3 with semimetal nanoinclusions, *Adv. Energy Mater.*, **1**(6), pp. 1–7.

87. Zhao, X. B., Ji, X. H., Zhang, Y. H., Zhu, T. J., Tu, J. P., Zhang, X. B. (2005). Bismuth telluride nanotubes and the effects on the thermoelectric properties of nanotube-containing nanocomposites, *Appl. Phys. Lett.*, **86**(6), p. 062111.

88. Chen, C. L., Chen, Y. Y., Lin, S. J., Ho, J. C., Lee, P. C., Chen, C. D., Harutyunyan, S. R. (2010). Fabrication and characterization of electrodeposited bismuth telluride films and nanowires, *J. Phys. Chem. C*, **114**(8), pp. 3385–3389.

89. Toprak, M., Zhang, Y., Muhammed, M. (2003). Chemical alloying and characterization of nanocrystalline bismuth telluride, *Mater. Lett.*, **57**, pp. 3976–3982.

90. Kim, K. T., Koo, H. Y., Lee, G. G., Ha, G. H. (2012). Synthesis of alumina nanoparticle-embedded-bismuth telluride matrix thermoelectric composite powders, *Mater. Lett.*, **82**, pp. 141–144.

91. Chávez-Ángel, E., Reparaz, J. S., Gomis-Bresco, J., Wagner, M. R., Cuffe, J., Graczykowski, B., Shchepetov, H., Jiang, A., Prunnila, M., Ahopelto, J., Alzina, F., Sotomayor Torres, C. M. (2014). Reduction of the thermal conductivity in free-standing silicon nano-membranes investigated by non-invasive Raman thermometry, *APL Mater.*, **2**(1), p. 012113.

92. Liang, B., Song, Z., Wang, M., Wang, L., Jiang, W. (2013). Fabrication and thermoelectric properties of graphene/Bi_2Te_3 composite materials, *J. Nanomater.*, **2013**, Article ID 210767.

93. Goldsmid, H. J. (2014). Bismuth telluride and its alloys as materials for thermoelectric generation, *Materials*, **7**(4), pp. 2577–2592.

94. Keshavarz, M. K., Vasilevskiy, D., Masut, R. A., Turenne, S. (2013). p-Type bismuth telluride-based composite thermoelectric materials produced by mechanical alloying and hot extrusion, *J. Electron. Mater.*, **42**(7), pp. 1429–1435.

95. Deng, Y., Nan, C. W., Wei, G. D., Guo, L., Lin, Y. H. (2003). Organic-assisted growth of bismuth telluride nanocrystals, *Chem. Phys. Lett.*, **374**, pp. 410–415.

96. Liao, C. N., She, T. H. (2007). Preparation of bismuth telluride thin films through interfacial reaction, *Thin Solid Films*, **515**, pp. 8059–8064.

97. Sokolova, O. B., Skipidarova, S. Y., Duvankova, N. I., Shabunina, G. G. (2004). Chemical reactions on the Bi_2Te_3-Bi_2Se_3 section in the process of crystal growth, *J. Cryst. Growth*, **262**, pp. 442–448.

98. Kim, K. T., Ha, G. H. (2013). Fabrication and enhanced thermoelectric properties of alumina nanoparticle-dispersed $Bi_{0.5}Sb_{1.5}Te_3$ matrix composites, *J. Nanomater.*, **2013**, Article ID 821657.

99. Gothard, N., Wilks, G., Tritt, T. M., Apowart, J. E. (2010). Effect of processing route on the microstructure and thermoelectric properties of bismuth telluride-based alloys, *J. Electron. Mater.*, **39**(9), pp. 1909–1913.

100. Thiebaud, L., Legeai, S., Ghanbaja, J., Stein, N. (2018). Synthesis of Te-Bi core-shell nanowires by two-step electrodeposition in ionic liquids, *Electrochem. Comm.*, **86**, pp. 30–33.

101. Kim, J., Lee, J. Y., Lim, J. H., Myung, N. V. (2016). Optimization of thermoelectric properties of p-type $AgSbTe_2$ thin films via electrochemical synthesis, *Electrochim. Acta*, **196**, pp. 579–586.

102. Kwak, M. H., Kang, S. B., Kim, J. H., Lee, J., Lee, S. M., Kim, W. J., Moon, S. E. (2017). Aerosol deposition of thermoelectric p-type $Bi_{0.5}Sb_{1.5}Te_3$ and n-type $Bi_2Te_{2.7}Se_{0.3}$ thick films, *J. Ceram. Proc. Res.*, **18** (10), pp. 731–734.

103. Chung, D. D. L. (2017). Processing-structure-property relationships of continuous carbon fiber polymer-matrix composites, *Mater. Sci. Eng. R*, **113**, pp. 1–29.

104. Mahmoud, L., Alhwarai, M., Samad, Y. A., Mohammad, B., Laio, K., Elnaggar, I. (2015). Characterization of a graphene-based thermoelectric generator using a cost-effective fabrication process, *Energy Procedia*, **75**, pp. 615–620.

105. Lee, S., Kim, J., Ku, B. C., Kim, J., Joh, H. I. (2012). Structural evolution of polyacrylonitrile fibers in stabilization and carbonization, *Adv. Chem. Eng. Sci.*, **2**, pp. 275–282.

106. Saha, B., Schatz, G. C. (2012). Carbonization in polyacrylonitrile (PAN) based carbon fibers studied by ReaxFF molecular dynamics simulations, *J. Phys. Chem. B*, **116**, pp. 4684–4692.

107. Ma, Q., Gao, A., Tong, Y., Zhang, Z. (2016). The densification mechanism of polyacrylonitrile carbon fibers during carbonization, *New Carbon Mater.*, **31**(5), pp. 550–554.

108. Hameed, N., Sharp, J., Nunna, S., Creighton, C., Magniez, K., Jyotishkumar, P., Salim, N. V., Fox, B. (2016). Structural transformation of polyacrylonitrile fibers during stabilization and low temperature carbonization, *Polym. Degrad. Stab.*, **128**, pp. 39–45.

109. Liu, J., Wang, P. H., Li, R. Y. (1994). Continuous carbonization of polyacrylonitrile-based oxidized fibers: aspects on mechanical properties and morphological structure, *J. Appl. Polym. Sci.*, **52**, pp. 945–950.

110. Wang, H., Zhang, X., Zhang, Y., Cheng, N., Yu, T., Yang, Y., Yang, G. (2016). Study of carbonization behavior of polyacrylonitrile/tin salt as anode material for lithium-ion batteries, *J. Appl. Polym. Sci.*, **133**, 43914.

111. Sun, J., Wu, G., Wang, Q. (2005). The effects of carbonization temperature on the properties and structure of PAN-based activated carbon hollow fiber, *J. Appl. Polym. Sci.*, **97**, pp. 2155–2160.

112. Rahaman, M. S. A., Ismail, A. F., Mustafa, A. (2007). A review of heat treatment on polyacrylonitrile fiber, *Polym. Degrad. Stab.*, **92**, pp. 1421–1432.

113. Gan, Y. X., Chen, A. D., Gan, J. B., Anderson, K. R. (2018). Electrohydrodynamic casting bismuth telluride micro particle loaded carbon nanofiber composite material with multiple sensing functions, *J. Micro Nano-Manuf.*, **6**(1), pp. 011005-1–011005-9.

Index

PGMO 08/23/2018